RICE BREEDING AND GENETICS

About the Author

Supriyo Chakraborty (*b.* 1968) received his B.Sc. (Agri.) degree with"Distinction" in 1990 and M.Sc. (Agri.) degree in Plant Breeding and Genetics in 1992 from Assam Agricultural University, Jorhat. He obtained his Ph.D. degree in Life Sciences (Botany) in 1998 from Dibrugarh University, Dibrugarh. He has throughout a brilliant academic career.

He qualified NET (National Eligibility Test) for lecturership in Life Sciences conducted by CSIR, India in 1992. He also qualified NET for lecturership in Plant Breeding and in Genetics conducted by ICAR in 1996 and 1997 respectively. Dr. Chakraborty has about 8 years of research experience in breeding of rice, tea and various pulse crops. He has published more than 25 research papers in scientific journals. He is a member of several scientific societies. Prior to joining the present post of Scientist (Plant Breeding and Genetics) in the rank of Assistant Professor at Regional Agricultural Research Station, Assam Agricultural University, Shillongani, P.O. Nagaon 782 001, Assam, he served as JRF in the Department of Biotechnology, AAU, Jorhat and as SRA in the Ministry of Textiles, Government of India.

Author's permanent address: C/o Annada Medical Hall, P.O. Mariani-785634, Assam, India.

RICE BREEDING AND GENETICS

SUPRIYO CHAKRABORTY, Ph.D.
Assistant Professor
Department of Plant Breeding and Genetics
Assam Agricultural University
Jorhat (India)

CONCEPT PUBLISHING COMPANY, NEW DELHI-110059

ISBN 81-7022-874-3 (HB)

First Published 2001

Printed and Published by
Ashok Kumar Mittal
Concept Publishing Company
A/15-16, Commercial Block, Mohan Garden
New Delhi-110059 (India)

Phones: 5648039, 5649024
Fax: 091–(11)–5648053
Email: publishing@conceptpub.com

DEDICATED
TO
MY TEACHER,

PROF. G.N. HAZARIKA

Foreword

Over the last four decades, rice breeding has seen a sea-change in the avenues of research aimed at yield and quality improvement. Present day rice breeding with its high efficiency derives strength from the works on genetics. Since the development of high yielding rice varieties in the sixties by introducing dwarfing genes, no significant yield increase in rice has been achieved. This is attributed to the concentrated efforts of rice breeders in stabilising rice yield by building up disease and insect pest resistance in the genetic background of high yielding rice varieties. No serious effort was made since 1970s by the rice breeders to enhance the genetic yield potential of rice directly.

Ever growing human populations in Asia, Africa and Latin America have forced the rice breeders to tailor the rice plant for increasing yield through heterosis breeding, wide hybridization, ideotype breeding, biotechnology and genetic engineering. Considerable research throughout the world has been undertaken in these and allied areas resulting in the accumulation of valuable research information. These informations quite often stay bound in research journals and do not reach the students. As a result, most of the students remain unaware of the recent developments in rice breeding. In this book, Dr. Supriyo Chakraborty has attempted to provide the latest developments in rice breeding and genetics in simple and intelligible language . The beauty of this book is that few chapters have been devoted to rice molecular breeding. Chapters like "Hybrid rice breeding", "Wide hybridization in rice", "Genetic markers in rice" and "Quantitative trait loci" will help the students in understanding the complicated topics in a simpler way and in undertaking further research easily for rice genetic improvement. In addition, the chapter on "Rice biotechnology and genetic engineering" provides information on the present status of research in that area. Besides these, chapters on "Early rice research in India", "Chromosome structure" and "Aneuploidy" will be of great help to the beginners.

The undergraduate as well as the postgraduate students of agriculture in general and of the Plant Breeding and Genetics discipline in particular will find this text book very useful in updating their information. This text book also bridges the gap of not yet having a book with the latest information on rice breeding and genetics. Dr. Norman E. Borlaug, the eminent plant breeder of Mexico and Nobel laureate once mentioned: "Plants speak to men but only in whisper ; their voice can be heard only by those who remain close to them." In order to develop this intimacy with the rice plant, it is very essential to understand rice plant at molecular level. The present text book is a step towards bringing the readers to that goal.

I personally appreciate Dr. Supriyo Chakraborty for his maiden attempt at writing such a valuable book on rice breeding and genetics with latest information in simple language. I hope this book will find wide acceptance in the universities, colleges and rice research institutions of India and other south-east Asian countries.

Prof. A.N. Mukhopadhyay
Ph.D., F.A.A.Sc.
VICE-CHANCELLOR
Assam Agricultural University
Jorhat

12 January, 2000
Jorhat

Introduction

In the last decade, the science of plant breeding and genetics has ushered in the era of electronic speed for increasing global food production. In this noble endeavour, biotechnology and genetic engineering, still being in their infancy, have already begun contributing. Rice breeders and geneticists also are not lagging behind in harnessing new tools of biotechnology and genetic engineering. They have been exploring various conventional and non-conventional research avenues to develop rice varieties with higher yield, improved quality, better biotic and abiotic stress tolerance for combating the hazards of rice production. A pile of information on rice breeding and genetics has been generated by the scientists over the previous decades.

The undergraduate and the postgraduate students of Agricultural Universities should be abreast of the latest developments in rice genetic improvements. Nowadays, basic knowledge of genetic markers, QTLs, ideotypes, RFLPs and other genetic engineering tools is essential in undertaking research for genetic manipulations. Students as well as researchers have had a long-felt need for a textbook on rice breeding and genetics containing latest information. I congratulate Dr. Supriyo Chakraborty on his maiden attempt to compile and present a part of the most useful latest developments on rice breeding and genetics in this book. The book is hence a timely contribution.

I wish that the students as well as the rice researchers will find this book very practical and useful.

Prof. A. K. Pathak
Director of Research (Agri.)
Assam Agricultural University,
Jorhat

Preface

Plant breeding is the genetic improvement of crop plants in relation to their economic use for mankind. Plant breeding is offered as a subject at the undergraduate and the postgraduate levels in all the Indian Agricultural Universities. It is also offered at graduate level in biology and postgraduate level in botany in conventional universities. Rice is one of the major crops covered under the subject Plant Breeding at all the universities. Rice breeding aims at evolving high yielding and quality rice varieties for feeding the ever-growing human population.

Present day rice breeding with its high efficiency derives strength from the works on genetics. Knowledge of genetics (both heredity and variation) is a pre-requisite for developing desired varieties through breeding programmes. Rice breeders across countries have been engaged in developing improved varieties by applying the knowledge of genetics. There are a few books especially by foreign authors with elaborate discussion on particular topics of rice breeding and genetics. However, there is no single book which precisely covers all aspects of rice breeding and genetics.

This book entitled "Rice Breeding and Genetics" covers in one compact volume the latest developments in rice breeding and genetics in addition to several fundamental topics of general interest. The language used in the book is simple and easily understandable. The presentation is pointwise and step by step with recent Indian examples, wherever applicable. The references are given at the end of each chapter.

It is undoubtedly an impossible task to compile all the information, gathered over years, on rice breeding and genetics in a single volume. However, I have endeavoured in this book to acquaint the students and research workers with the latest developments in rice breeding and genetics precisely covering the syllabus.

History not only serves as a link between past and present but also shapes the future. To abide by the convention that history always comes first, this book also begins with the chapter" Early rice research in India.Several chapters of general interest namely "Rice ecosystems", "Rice photoperiodic response", "Development of hybrid rice in India", "Global status of hybrid rice research" and "Rice seed production and testing" etc. have been included as a source of basic information for the students.

In order to keep the undergraduate students of agricultural and conventional universities abreast with the latest developments in rice genetic improvements, chapters like "Rice chromosome structure", "Rice aneuploidy", "Hybrid rice breeding", "Wide hybridization in rice", "Rice ideotype breeding", "Rice yield improvement" and "Rice mutation breeding in India" have been compiled in this book. Besides these, a few important chapters like "Genetic markers in rice", "Quantitative trait loci (QTLs)", " RFLP technique" and "Enhancing genetic yield potential in rice" included in this book will be of great help to the postgraduate students interested in pursuing research for genetic manipulations at molecular level. These topics either did not find any place or received inadequate treatment in most of the previously published textbooks on Plant Breeding or Rice Breeding. The last chapter on "Rice biotechnology and genetic engineering" provides information in a nutshell on the present status of research in this area. The teachers and the researchers will find the chapters of this book very useful to update their knowledge.

This book also bridges the gap of not having a textbook with the latest information on rice breeding and genetics. Candidates appearing for the competitive examinations such as ARS, NET and SRF examination of ICAR may use this book as a reference. In my opinion, no textbook can claim originality either in content or in presentation. The present textbook is no exception and it draws heavily from several recent quality books and rice literatures published in various national and international journals.

All omissions, errors and deficiencies in this textbook are my own and none else should be credited for them. I am sure that many worthwhile chapters have been left out of this book mainly due to my ignorance. I hope the readers will not hesitate to interact with me in order to improve this text book by suggesting inclusion of other worthwhile chapters. I shall remain ever grateful to the readers for their suggestions and constructive criticisms.

I am very grateful to Dr. I.D. Singh, Commonwealth Consultant to the Tea Industry of Sri Lanka. My long association with him has instilled in me the confidence required for writing a book and also the habit of spending long hours with the work at hand.

I wish to appreciate Dr. B. Barman, Dr. C.R. Sarkar. Dr. B. Guha, Dr. P.K. Borua, Dr. P.K. Das, Dr. P.C. Sarmah, Dr. B.K. Borah, Dr. H.K. Borah, Dr. M.C. Debnath, Mr. V.S. Borah and other well-wishers who have encouraged me for the completion of this book. I have received great help from the Library of International Rice Research Institute (Philippines) which has mailed me several literatures on request.

I owe to my parents, parents-in-laws and relatives who have enthused me in writing this book. I extend my gratitude to Rohit, Bishnu, Tapan and Mousumi for their untiring efforts for typing this book. Despite all these encouragements, I feel myself very fortunate in having a wife who understands and helps me to overcome anxiety in various matters. I am greatly indebted to my wife Sumita, who has been patient enough during the long hours I spent in preparing this book notwithstanding her busy work schedule at home as a housewife and in school as a teacher.

September, 1999 SUPRIYO CHAKRABORTY

RARS, AAU,
Nagaon - 782 001 (Assam)

Contents

1

Early Rice Research in India

World Production Scenario

Rice is the most important staple foodgrain in the world. It is cultivated in 148 million hectares (ca.) worldwide which accounts for nearly 10% of the world's arable land. About 95% of the world's rice is produced in developing countries primarily in Asia which is home to 59% of the world's population. Asian farmers harvest nearly 92% of global rice produce. Rice provides 55 to 85% of the total calories consumed per adult in developing countries like Cambodia, Myanmar, Bangladesh, Thailand, Laos, Vietnam and Indonesia. In most of the African and Latin American countries rice meets less than 10% of the total calories consumed whereas in Guinea, Guyana, Madagascar and Liberia about 25 to 50% of the calories eaten come from rice.

Poverty and low per capita income are found in most of the rice producing countries. Nearly 12 countries out of 23 that produce more than 1 million tonne of rice have a per capita income less than US $500. It has been estimated that by 2025 A.D. about 758 million tonnes of rice will be required to feed the global population as against the present production of 520 million tonnes.

World Trade

In Asia most of the farmers grow rice on small farms of 1 to 3 hectares to meet family requirements and only a very small proportion goes to local market. In Brazil the farmers grow rice on commercial farms of more than 50 hectares. Despite the large global production, only 4% of the total world produce goes to international rice trade and that too only in quality scented rice like Basmati. Major exporters of rice to the world market are USA (19%), Thailand (36%), Vietnam (10%), Pakistan (7%),. India and China each account for 1-3% of the world's market.

The driving forces behind the increased production of rice over the last 30 years were the genetic improvement of varieties and development of production technology. The first modern variety of rice IR8 was released by IRRI, Philippines in 1966 and since then quite a large number of IR parented rice varieties have been released. Farmers harvest 2 to 3 times higher yield from the modern varieties than from traditional rice varieties. Yield stabilising factors like greater pest and disease resistance, shorter crop duration have been incorporated into the modern rice varieties.

Green Revolution and Rice

The development of semidwarf rice varieties ushered green revolution in India. Rice varieties grown in India mostly belong to the sub-species indica. The Chinese scientists came across a spontaneous mutant in Dee-geo-woo-gen, a dwarf early maturing variety of japonica sub-species from Taiwan. The evolution of TN 1 (Taichung Native 1) by crossing a tall indica variety "Tsai Yuen Chung" with "Deo-geo-woo-gen" in Taiwan and IR8 (Miracle rice) from a cross between Peta, a tall Philippine indica variety and "Deo-geo-woo-gen" at International Rice Research Institute, Los Banos, Philippines in 1960s brought a real revolution in rice production. These semidwarf indica rice varieties are fertilizer responsive, high yielding, lodging resistant and photoinsensitive. These two indica varieties could dispel the notion that only japonica varieties of rice are capable of responding to high fertilizer. In 1965-66 both the varieties were introduced in India and subsequently recommended for commercial cultivation. Photoinsensitivity has been a boon for extending rice cultivation in non-traditional areas like Punjab and Haryana.

Rice Varietal Improvement in India

Varietal improvement work in India was initiated in a systematic manner as early as 1911 (Seetharaman, 1978). Paddy specialist in Madras State and economic botanists in Bengal and Bombay were assigned the task of varietal improvement. Several rice varieties were developed through pure line selection from heterogeneous populations each comprising a mixture of homozygous genotypes at Coimbatore in Madras, Chinsurah in Bengal, Karjat in Bombay and in many research institutions of the country. Then the japonica varieties were tested for their direct introduction in India but were found to be unsuccessful in several parts of the country. In the mid 1940's several varieties like 'Ch 2', 'Ch 10', 'Ch 45', 'Ch 988' and 'Ch 1039' from China, 'Shinei' from Japan, 'Tainan 3' and 'Kaohsiung 22' from Taiwan, 'Mahsuri' from Malaysia etc. were introduced to India for cultivation. Simultaneously, state-wise testing of varieties led to the spreading of 'Ptb10' from Kerala and 'Mtu 15' from Andhra Pradesh to Orissa, 'Bam 12' from Tamil Nadu to Punjab, 'N136' from Uttar Pradesh and 'FR 43 B' from Orissa to Bihar.

Hybridization Work

Hybridization work in rice was started at a later stage in different states of the country and it led to the development of several varieties such as 'Co1', 'Mtu 16' and blast resistant strains namely 'Adt 25', 'Co 5', 'Co 26' and 'Co 31'. Nearly 49 varieties developed through hybridization included at least one indigenous parent. Later in 1960's, a javanica variety 'Urang Urangan 89' was used as one of the parents in crosses and resulted in the development of a late-maturing and superfine grain variety 'CR 1014' at Central Rice Research Institute, Cuttack, Orissa. A short-statured variety 'Baok' from Indonesia was also included in the programme to evolve short-statured varieties with good yield potential. Breeding projects were initiated using 'TKM 6' as one of the parents to develop a high yielding variety with tolerance to stem borer.

Establishment of Central Rice Research Institute (CRRI)

The Central Rice Research Institute at Cuttack in Orissa was established by the Government of India in 1946 to conduct research on all aspects of rice production and transferred to ICAR in 1966. The CRRI co-ordinates :

1. Collection, maintenance and evaluation of rice germplasm throughout the country.
2. Rice improvement work in India through the All India Co-ordinated Rice Improvement Project (AICRIP).

Besides this, the CRRI carries out research on rice genetics, breeding, disease and pest problems, economics, agronomy, meteorology and other rice related problems.

Mutation Study

Dr. K. Ramiah, a well known rice breeder and the first director of CRRI, attempted mutation breeding in rice in 1936 and isolated a number of short-statured mutants. The mutants were not superior to the then existing improved varieties under low management. In 1969, a high yielding variety 'Jagannath' obtained through mutation in variety 'T 141' was released. The effect of some chemical mutagens on few rice varieties was studied at Central Rice Research Institute, Cuttack in 1966. Some short-statured productive mutants were isolated in 'Saturn' and 'Tainan3' despite a large number of morphological variants.

Research work carried out till 1950s resulted in the release of a large number of varieties and none of them could exceed the yield level of 4 tonnes/ha. The yield of the improved varieties remained almost static till early 1960s. In 1964, 'Taichung Native 1' (TN1) was introduced to India and this was followed by 'IR 8' in 1965 developed at IRRI, Philippines. Both 'TN 1' and 'IR8' were short-statured, lodging-resistant and nitrogen-responsive varieties. These varieties were used as donors in several hybridization programmes and as a result, 10 varieties had been released up to 1970 by the Central Variety Release Committee.

Table 1.1: List of Donors in HYV programme in India

Characters	Varieties
Earliness	N 22, Blue Belle, Adt 27
Drought tolerance	(Taichung 65 X TN 1), Lal Nakanda 47, Mtu 17
Cold tolerance	Dunghanshaly
Saline tolerance	Getu, Dasal, Damodar, Bokra.
Fine grain	T9, GEB24, Basmati370, T141
Resistance to	
- blast	Te-tep, Taduksn, Zenith, Co4, S67
- bacterial blight	BJ-1, Early Prolific, Wasei- Aikoku3, Malagkit - Sung Sung
- stem rot	Basmati 370, Bara 62
- tungro virus	Kataribhog
Tolerance to	
- Stem borer	TKM 6, W 1263, CB1, CB 2
- gall midge	Ptb18, Ptb21, Siam 29, Leaung 152, Peykeo E 53, Peykeo P 129, W 1263, CR56-17
- leaf hopper (green)	Vijaya, Pankhari 203, Asd 7, IR8
- plant hopper (brown)	Mudgo, Asd 7, Ptb 18

Courtesy : Rice Production Manual, ICAR (1978), p. 28.

Promising Rice Varieties Released in India in the Past

1. **Bala :** Bala is an early maturing variety which takes 90 days from seed to seed under direct seeded condition and about 100 days under transplanted condition. The plants are short-statured (87 cm) and have a good nitrogen response with a tendency to lodge at very high nitrogen levels. Grains are short and bold. It is tolerant to drought, insensitive to photoperiod and slightly thermosensitive. It has seed dormancy.
   Bala is resistant to some races of blast but is susceptible to bacterial blight and some insect pests like gall midge, stem borer and leaf hopper.
2. **Cauvery :** It is an early maturing (105 days), short, bold grained variety and developed from a cross between 'TN 1' and

'TKM 6'. The plants are short-statured (88 cm). It is fertilizer responsive and has low tolerance to drought and phosphorus deficiency.

3. **Jamuna :** It is a selection from the backcross progenies of 'TN 1' x 'Basmati', short-statured (84 cm) with a maturity duration of 112 days. The grains are long, slender and non-scented. It is photoinsensitive and moderately resistant to some races of blast but susceptible to bacterial leaf blight, gall midge and stem borer.
4. **Sabarmati :** It is a selection from the back cross progenies of 'TN 1" x 'Basmati 370' with a maturity period of 118 days. The plants are 86 cm tall and photoinsensitive. The grains are medium slender and slightly scented. It is moderately resistant to some races of blast but susceptible to bacterial leaf blight, gall midge and stem borer. This variety is suitable for multiple cropping due to its earliness.
5. **Ratna :** It is a short-statured variety (83 cm) with long superfine grains, good milling quality, excellent cooking quality and developed from a cross between 'TKM 6' and 'IR 8'. It is the first high yielding variety having tolerance to stem-borer at the dead heart stage. It is moderately resistant to most races of blast but susceptible to bacterial blight and gall midge.
6. **Padma :** It was developed from the cross 'T 141' x 'TN 1'. The variety is short-statured (81 cm) with 120 days maturity duration. The grains are short, bold without white belly and the milling recovery is good. It is susceptible to gall midge, leaf hopper and stem borer but moderately resistant to blast. Padma is grown as a boro rice in West Bengal.
7. **Jaya :** It was evolved from a cross between 'TN 1' and 'T 141'. The plants are 83 cm tall with a maturity period of 130 days and have good tillering habit. The grains are coarse with white kernels. It is photoinsensitive, responsive to nitrogen and sensitive to low temperature. It is moderately resistant to blast, tolerant to bacterial leaf blight and susceptible to gall midge and stem borer.
8. **IR 8 :** It was evolved from a cross between 'Peta' and 'Dee-geo-woo-gen' in IRRI, Philippines and introduced to India in

1965. Plants have good early seedling vigour with erect leaves and late leaf senescence. It is highly responsive to nitrogen and resistant to lodging and several races of blast. It is tolerant to bacterial blight but susceptible to gall midge. Grains are long, bold with white kernels and have good cooking quality. IR 8 is photoperiod insensitive and considered the first high yielding indica rice variety adapted to tropical climates.

9. **IR 20 :** It was developed at IRRI, Philippines from a cross between 'IR 262-24-3' and 'TKM 6'. Plants are short-statured (100 cm) with a maturity duration of about 135 days. It is photoperiod insensitive and moderately responsive to nitrogen. The grains are medium slender and translucent. It is moderately resistant to bacterial blight and tungro virus.
10. **Sona :** It was evolved from the cross between 'GEB 24' and 'TN1'. It has long slender grains and a good cooking quality. Panicles are of hard threshing type.
11. **Pankaj :** Pankaj which was developed from a cross between 'Peta' and 'Tongkai Rotan' was introduced to India from IRRI, Philippines. The variety is photoperiod insensitive and responsive to nitrogen. The grains are long and bold. Pankaj is susceptible to a few races of blast and moderately susceptible to bacterial leaf blight. It is also susceptible to gall midge and stem borer. This variety is suitable for areas where long duration rice varieties are commonly grown.
12. **Jagannath :** It is a mutant rice variety obtained from T 141. The variety is late maturing type. Seeds are slender and possess dormancy. It is responsive to nitrogen and possesses resistance to lodging and suitable for low lying areas.

    Jagannath is susceptible to blast, bacterial leaf blight, stem borer and gall midge.

REFERENCE

Seetharaman, R. 1978. Varieties and regional adaptability. *Rice Production Manual,* ICAR, New Delhi, 24-37.

2

Rice Ecosystems

Introduction

Rice crop grows well in warm climate with adequate moisture. The principal rice producing countries lie between 14^{0}S latitudes and 24^{0}N latitudes. Rice is grown in tropics and subtropics from below the sea level upto an elevation of 2500 m or even higher. Rice is a water-loving plant. Based on the anatomical features of root and stem, it is sometimes described as a semi-aquatic plant (Shiv Raj, 1987). Yields of rice are quite low during the shortest and the coldest days of the year.

Rice Ecotypes

In India, three distinct ecotypes of rice are commonly found. These are :

i) lowland rice cultivated under flooded conditions;
ii) upland rice grown during monsoon period; and
iii) deep water rice or floating rice grown mostly in Assam and West Bengal.

Generally the yields of lowland rices are higher than upland rices and nearly 95 percent of the rice belongs to lowland type. Some rice varieties grow well under both lowland and upland conditions.

Deepwater rice is also known as bao rice in Assam. It is usually sown in March-April and harvested in late November or December.

Deepwater rice refers to those ecotypes which are planted in the crop field when the standing water for a certain period of time is more than 50 cm and when the water depth varies between 51 and 100 cm for more than half of the growth duration. On the other hand, floating rice refers to those deepwater varieties planted to the field where maximum water depth ranges from 1 m to 6 m for more than half of the growth duration.

In India, deepwater and floating rices are grown between 8°N and 30°N latitudes where the monthly average temperature varies from 15°C to 35°C, day length is 11 to 13 hours with average sun-shine of 4 to 10 hours and the rainfall is 100 cm to 600 cm during the crop life cycles. The characteristics desirable for deepwater rices are good seedling vigour, intermediate plant height, submergence tolerance when flooded, photoperiod sensitivity, elongation ability of the stem with rising water level and kneeing ability of the stem after the recession of flood water.

Criteria of a Rice Growing Region

A region for being suitable for rice crop must have some characteristics *e.g.*

(1) Relatively warm climate throughout the growing season with a mean temperature of 21°C or above.
(2) An adundant supply of water for proper irrigation.
(3) Soils that are fairly level and capable of holding water fairly well without too much seepage.
(4) Adequate surface drainage (Shiv Raj, 1987).

Rice Ecosystems

The ecosystems within which rice is grown are characterised by rainfall pattern, elevation, depth of flooding, drainage and adaptation of rice to these agro-ecological factors. Rice is usually grown under four ecosystems throughout the world.

1) **Upland Ecosystem :** Under this ecosystem, rice is direct seeded

in non-flooded, well-drained soil on level to steeply sloping field. Crop suffers from lack of moisture and inadequate nutrition. So yields of upland rice are generally very low. Upland rice is grown on about 19 million hectares of land worldwide. Upland rice ecosystem is dominant in Africa and South America but relatively less important in Asia.

(2) **Irrigated Ecosystem :** Rice is direct seeded or transplanted in puddled soil on levelled bunded field with water control in both dry and wet seasons, in summer in higher elevations and in dry season in the flood prone areas. The crop is heavily fertilised. Yield can reach as high as 5.0 tonnes/hectare in wet season and more than 8 tonnes/hectare in dry season. Irrigated rice is planted on 81 million hectares worldwide. It accounts for 55% of world's harvested rice area and contributes nearly 76 per cent to global rice production. Irrigated ecosystems are concentrated in semi-arid and sub-humid subtropics of Asia.

(3) **Rainfed Lowland Ecosystem:** Rainfed lowland rice is grown on 37 million hectares worldwide. Rice is transplanted or direct seeded in puddled soil on level to slightly sloping bunded fields with variable depths and duration of flooding depending on rainfall. Soils alternate from flooded to nonflooded. Rice yield varies depending on rainfall, cultivation practices and use of fertilisers. This ecosystem of rice is dominant in humid and subhumid tropics.

(4) **Flood Prone Ecosystem :** In this ecosystem rice is direct seeded or transplanted in rainy season on fields characterised by medium to very deep flooding (50 to more than 300 cm) from rivers or from tides in river mouth deltas. Soils cycle from dry to flooded and may have severe problems of salinity and toxicity. The crop grows as flood water rises and harvesting is done when flood water recedes. This ecosystem covers more than 10 million hectares land worldwide and predominantly exists in South and South-East Asia.

REFERENCE

Shiv Raj, A. 1987. *An Introduction to Physiology of Cereal Crops.* Oxford & IBH Publishing Co. Pvt. Ltd., New Delhi, 1- 40.

3

Rice Photoperiodic Response

Photoperiodic Response

The length of a day, defined as the interval between sunrise and sunset, is called photoperiod.

Photoperiodism is the response of the plant to photoperiod. Rice is basically a short-day plant. In other words, rice plant needs a long night for initiation of panicle primordia. When the rice plant is subjected to long photoperiods (*i.e.* long days), panicle primordia are initiated late or even may not develop at all (Yoshida, 1981).

During the early stages of growth, all varieties show a period of insensitivity to photoperiods ranging from 2 to 10 weeks. Short photoperiods of 4-12 hours generally hasten floral initiation and heading while long photoperiods of 13-24 hours tend to prolong the vegetative period. In India, flowering of summer varieties normally takes place when the seasonal day length is above 12 hours, while winter varieties flower when the day length is below 12 hours (Murneek and Whyte, 1948). About 8 hours of photoperiod is detrimental to early varieties, it reduces growth and delays panicle emergence and subsequently reduces grain yield (Shiv Raj, 1987).

The vegetative growth stage of rice is divided into two phases:

(a) **Basic Vegetative Phase (BVP) :** It is the minimum duration of vegetative growth required before the plant will initiate panicle primordium which can be triggered by short days. The length of BVP ranges from 10 to 63 days. No variety has been found to have BVP longer than 100 days. One simple way to estimate the BVP of a variety is to subtract 35 days from growth duration (*i.e.* from sowing to days to flower) at the optimum photoperiod (Yoshida, 1981). The optimum photoperiod of most rice varieties is about 9-10 hours (Yoshida, 1981) or 9-12.5 hours (Best, 1959) . Optimum photoperiod is the day length at which the duration from sowing to flowering is at a minimum.

BVP = Shortest growth duration – 35 days

(b) **Photoperiod-Sensitive Phase (PSP) :** PSP is that portion of the vegetative phase affected by day length. PSP is also called lag vegetative growth. After the BVP, the plant enters the PSP during which floral initiation can be triggered by short days. The effect of day length differs with varieties. In some varieties, it determines whether the plant will flower or not. The PSP of photoperiod insensitive varieties is less than 30 days; that of sensitive varieties, longer than 31 days.
The differences in the growth duration of photoperiod-sensitive or seasonal varieties result mainly from variations in the length of the PSP. The day length when the plant has finished its BVP, in turn, determines the variations in the length of the PSP (Vergara, 1983).

PSP = longest growth duration – shortest growth duration

Critical photoperiod (CP) and optimum photoperiod (OP) are the two terms commonly used to characterise the PSP. Critical photoperiod is the longest photoperiod at which the plant will flower or the photoperiod beyond which it cannot flower. The CP of most varieties varies from 12 to 14 hours. Generally, strongly photoperiod sensitive varieties do not produce panicle primordia or they take a very long time to form panicle if the day lengths are longer than certain critical duration. In the photoperiod-nonsensitive varieties, there is no critical photoperiod since flowering occurs at any day length. Optimum photoperiod has been discussed in the preceding paragraph for basic vegetative phase.

Varietal Reactions to Photoperiod

The sensitivity of rice plants to photoperiod differs among varieties. Under similar climates, classification of rice varieties using the terms early, medium, late, summer or winter rice etc. is useful but in a wide range of climates classification of rice varieties based on photoperiod reaction is more rational.

Vergara and Chang (1976) suggested the following practical classification:

Reaction to photoperiod	Characteristics
(1) Photoperiod insensitive	Very short PSP (less than 30 days) and BVP varying from short to long
(2) Weakly photosensitive	Marked increase in growth duration when photoperiod is longer than 12 hours; PSP may exceed 30 days but flowering occurs under any long photoperiod.
(3) Strongly photosensitive	Sharp increase in growth duration with increased photoperiod ; no flowering beyond critical photoperiod, BVP usually short (not more than 40 days).

Generally, the growth duration of the strongly photoperiod-sensitive varieties varies greatly depending on the month or season during which they are planted. In weakly photosensitive varieties, the growth duration varies slightly depending on the season or month during which they are planted. But in the case of photoperiod insensitive varieties, the growth duration remains more or less the same regardless of the season in which thay are planted.

A comprehensive review of the literature on the flowering effects of photoperiod on the rice plant was published by IRRI (Bull. no. 9). Shiv Raj (1987) drew the following general statements from that review :

(1) High photoperiod sensitivity in rice is controlled by one

or two dominant genes; the basic vegetative phase is controlled by two or three genes of cumulative but unequal effect. A short BVP is dominant to a long one.

(2) Rice is a short day plant. Nearly all varieties mature within a shorter time under a short photoperiod (about 10 hours) but the degree of sensitivity varies greatly among varieties.

(3) Young leaves are more receptive to the photoperiodic stimulus than old leaves; even leaf sheaths may receive the stimulus.

(4) The photoperiodic stimulus is not translocated from one tiller to another.

(5) The minimum number of photo-inductive cycles necessary to initiate the panicle primordium varies from 5 to 24. It differs not only with varieties but also with the photoperiod. An appropriate photoperiod in 24 hours cycle is called one photo inductive cycle.

(6) The response of the rice plant to photoperiod is governed by the length of the longest continuous dark period. Very-short light breaks (3-15 minutes) during the dark period may markedly affect the time of heading in some varieties.

(7) For supplementary illumination or light breaks, red light is the most effective.

(8) Full day light is not necessary to prevent panicle initiation in strongly photoperiod-sensitive varieties. These varieties may react to light intensities as low as 1 to 100 lux. These intensities are just above the light of full moon. Twilight, therefore, may contribute significantly to the length of the photoperiod under natural day light conditions (Vergara, 1983).

(9) Not only the panicle initiation but also the development processes between initiation and anthesis may also be affected by the prevailing photoperiod.

(10) The BVP, PSP, critical photoperiod and optimum photoperiod differ among varieties. The variations in each are largely responsible for the differences in flowering response to photoperiod among varieties.

(11) The duration of the BVP ranges from 10 to 63 days while

the PSP can be extended to as long as 12 years (actually recorded). However, experiments are usually ended after 200 days and the concerned variety receives a PSP value of 200.

(12) For photosensitive varieties, the optimum photoperiod is about 10 hours while the critical photoperiod ranges from 13 to 14 hours.

Utility of Photoperiod Insensitive or Nonseasonal Varieties

Photoperiod insensitive varieties can be grown at any time of the year. These varieties, therefore, give the farmer more flexibility in crop planning and fitting them in multiple cropping systems. Very early maturing photoperiod insensitive varieties enable farmers to have two crops in a year. In the case of photoperiod insensitive varieties, introduction of varieties from one latitude to another is easier. Photoperiod insensitivity has expedited the spread of high yielding varieties in the tropics (Vergara, 1983).

Utility of Photoperiod Sensitive Varieties

Under rainfed rice cultivation, photoperiod sensitive varieties are desirable particularly when the onset of rainy season is considerably delayed. In most of tropical Asia, the beginning of rainy season is unpredictable and varies from year to year but the end of the season is fairly fixed. Photoperiod sensitive varieties, when planted late, may still mature at its usual time regardless of shortened growing periods. In such situations, if photoperiod insensitive varieties are grown, they often do not reach maturity due to shortened growing period and occasionally suffer from water stress at later growth stages after the rain ceases.

Traditional varieties grown in the tropics are tall, photoperiod sensitive and have long growing periods. Under high doses of fertilizer nitrogen, these varieties grow very tall and usually lodge at later growth stages reducing the yield. If photoperiod sensitive varieties are grown later than usual, they remain short and hence possess increased lodging resistance and give more yields. Floating rice varieties with long growth period are photoperiod sensitive. So far, no tropical variety with long growth period has been identified to be photoperiod insensitive (Vergara and Chang, 1976).

Phytochrome in Photoperiodic Response

In a short day plant like rice, a brief exposure to red light during critical photoperiod inhibits flowering. This inhibitory effect on flowering can be reversed by a subsequent exposure to far red light. This inhibition of flowering involves the activity of a photoreceptor called phytochrome. This photoreceptor exists in two forms *viz.*

(a) red light absorbing form designated as Pr
(b) far-red light absorbing form designated as Pfr.

These two forms are interconvertible. Each occurs in dimeric form (Pr/Pr or Pfr/Pfr). In the presence of red light (660-670 nm), Pr is converted to Pfr. When Pfr absorbs far-red light (725-735 nm), it is converted into Pr. Under darkness Pfr is gradually converted to Pr.

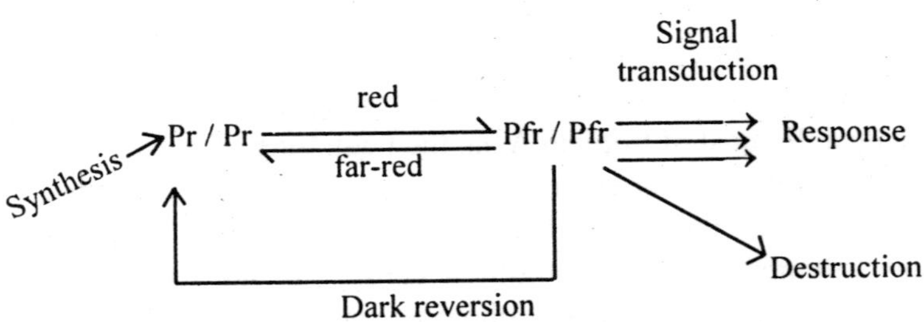

Fig. 3.1

Far-red light includes those wavelengths just longer than the red, covering approximately the range 700 to 800 nm but maximum absorption occurs at 730 nm wavelength (Salisbury and Ross, 1986). For red light maximum absorption by phytochrome pigment takes place at 666 nm wavelength. Chemically, phytochrome is a protein. One prosthetic group (an organic group or metal ion attached to a protein) called a chromophore which is an open chain tetrapyrrole is attached to the protein. The chromophore absorbs the light that induces phytochrome responses.

Sunlight primarily acts as a source of red light. During day time, Pfr is accumulated in the plant and flowering is inhibited. During critical photoperiod in rice, Pfr gradually changes to Pr which causes flowering. Phytochrome is present in most of the organs of rice plant including the roots.

REFERENCES

Best, R., 1959. Photoperiodism in rice. *Field Crop Abstract,* 12:85-93.

Murneek, A.E. and Whyte, R.O. 1948. Vernalisation and photoperiodism. *Chron Bot.* 1-196

Salisbury, F.B. and Ross, C.W. 1986. *Plant Physiology.* CBS Publishers and Distributors, New Delhi.

Shiv Raj, A. 1987. *An introduction to Physiology of Cereal Crops.* Oxford & IBH Publishing Co. Pvt. Ltd., New Delhi, 1-40.

Vergara, B.S. and Chang, T.T. 1976. The flowering response of the rice plant to photoperiod; a review of the literature. Technical Bulletin Number 8, IRRI, 75.

Vergara, B.S. 1983 . Plant growth and development. *Rice Production Manual.* University of Philippines and IRRI, 38-58.

Yoshida, S. 1981. *Fundamentals of Rice Crop Science.* IRRI, Philippines.

4

Rice Chromosome Structure

Introduction

Rice belongs to the genus *Oryza* which has about 23 species. Of these, two cultivated species are (i) *Oryza sativa* which is grown world wide and (ii) *Oryza glaberrima* which has limited area under cultivation in Africa. The cultivated varieties of *Oryza sativa* are classified into three groups *viz.* Indica, Japonica and Javanica, primarily on the basis of morphological character and geographic adaptation. The indicas include varieties from Southern China, Taiwan, India, Bangladesh, Pakistan, Nepal, Myanmar and Sri Lanka. The japonicas are commonly found in Japan, Korea and Northern China and these varieties are basically adapted to temperate climate. The javanica varieties are commonly found in Indonesia. The javanicas are intermediate between indicas and japonicas and are considered by some rice workers as the variants of the indica varieties (Chang, 1964).

Chromosome Number, Sporogenesis and Gametogenesis

The chromosome number of rice (*Oryza sativa* L.) was reported as 2n = 2x = 24 by Kuwada (1910) from the first study on rice microsporogenesis, megasporogenesis and mitosis. Microporogenesis (or megasporogenesis) is the formation of haploid microspore (or megaspore) from the diplod microspore (or megaspore) mother cell through meiosis. The microspore (pollen grain) undergoes two mitotic divisions to produce the microgametophyte containing two sperm nuclei and one tube nucleus and the process is called microgametogenesis (Fig. 4.1). On the other hand, the megaspore present in the ovary undergoes three mitotic divisions to produce an embryosac with eight haploid nuclei and the process is called megagametogenesis (Fig. 4.2) Various workers reported the chromosome number of both the cultivated species as 24 or 48 (Khush and Singh, 1991).

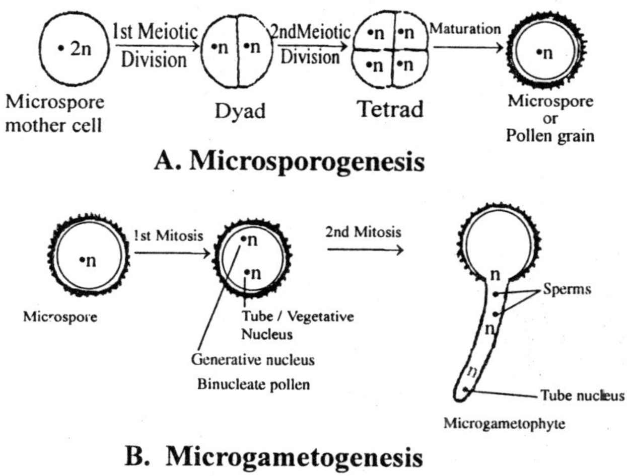

Fig. 4.1

Microsporogenesis and microgametogenesis occur in anthers. Megasporogenesis and megagametogenesis occur in ovary. One of the two sperms of the microgametophyte fuses with the egg cell to produce a diploid (2n) zygote. This process is called fertilization. The remaining sperm nucleus fuses with the diploid secondary nucleus (fusion product of two polar nuclei) of embryosac to form the triploid (3n) primary endosperm and the process is called triple fusion.

Genome Constitution

Genome is the basic set of chromosomes. In rice genome consists of 12 chromosomes which are different from each other in morphology and gene content. The chromosomes of a genome do

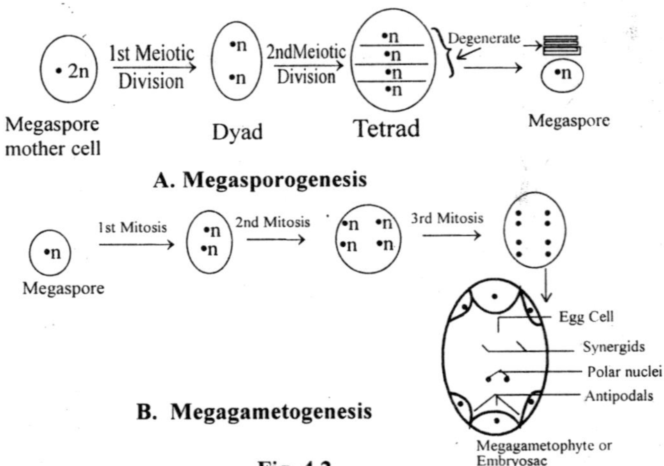

Fig. 4.2

not pair and hence produce no recombinants. Each chromosome of a genome forms one linkage group. In rice there are 12 linkage groups.

The genomic constitution of the two cultivated species *i.e. O. sativa* and *O. glaberrima* is AA and those of the wild species include AA, BB, CC, BBCC, CCDD, EE and FF. The genomes of some wild species have not been analysed as yet. The genomic constitutions and chromosome numbers of different species of *Oryza* are presented in Table 4.1 (Khush and Singh, 1991).

Genome Analysis

For some wild species of genus *Oryza*, only the chromosome numbers are known. The genomic analysis of some species has not been completed as yet. Both the cultivated species, *O. sativa* and *O. glaberrima* as well as related and wild species such as *O. nivera*, *O. rufipogon*, *O. breviligulata* and O. *longistaminata* have the AA genome. These species can be easily crossed with each other and they share a common gene pool. Their F_1 hybrids have normal chromosome pairing but show varying levels of sterility (Khush and Singh, 1991).

Table 4.1: Genomic constitutions and chromosome numbers of different species of *Oryza*

No.	Species	Genome	2n
1.	*O. sativa* L.	AA	24
2.	*O. glaberrima* Steud	A^gA^g	24
3.	*O. nivara* Sharma et Shastry	AA	24
4.	*O. perennis* Moench		
	Asiatic (subsp. *rufipogon*)	AA	24
	American (subsp. *cubensis* Ekman)	$A^{cu}A^{cu}$	24
	African (subsp. *barthii* A. Chev)	A^bA^b	24
5.	*O. breviligulata* A. Chev et Roehr	A^gA^g	24
6.	*O. longistaminata* A. Chev et Roehr	A^gA^g	24
7.	*O. punctata* Kotschy ex Steud	BB, BBCC	24, 48
8.	*O. minuta* J. S. Presel ex. C.B. Presl	BBCC	48
9.	*O. officinalis* Wall ex Watt	CC	24
10.	*O. eichengeri* A. Peter	CC, BBCC	24, 48
11.	*O. Latifolia* Desv.	CCDD	48
12.	*O. alta* Swallen	CCDD	48
13.	*O. grandiglumis* (Doell) Prod.	CCDD	48
14.	*O. australiensis* Domin.	EE	24
15.	*O. brachyantha* A. Chev et Roehr	FF	24
16.	*O. perrieri* A. Camus	Unknown	24
17.	*O. tisseranti* A. Chev.	Unknown	24
18.	*O. schlechteri* Pilger	Unknown	24
19.	*O. meyeriana* (Zoll. et. Mor. ex Steud) Baill	Unknown	24
20.	*O. granulata* Nees et Arn ex. Watt	Unknown	24
21.	*O. ridleyi* Hook. f	Unknown	48
22.	*O. longiglumis* Jansen	Unknown	48
23.	*O. indandamanica* Ellis	Unknown	24

Oryza genomes BB, CC, DD, EE and FF are quite distinct and very little intergenomic chromosome pairing is observed among them (Hu, 1970 ; Ogawa and Katayama, 1973; Katayama and Ogawa, 1974; Li *et al.*, 1961).

Numerous hybrids between the AA genome on the one hand and BB, CC, EE, FF, BBCC, and CCDD genome species on the other have been studied (Nezu *et al.*, 1960; Li *et al.*, 1962 and Katayama, 1966). In these intergenomic hybrid, only 0 to 5 bivalents are observed. But Nowick (1986) observed 10-18 bivalents in the hybrid between *O. sativa* (cultivar Mars) and *O. latifolia*. She postulated the presence of a pairing promoter gene in the A genome of *O. sativa*. It is necessary to identify the other genotypes of *O. sativa* which promote intergenomic chromosome pairing. These will be useful in obtaining recombinants between the cultivated species and the wild species with BB, CC, EE, FF, BBCC and CCDD genomes (Khush and Singh, 1991).

Somatic Karyotype

Karyotype is the general morphology of the somatic chromosome complement of a species. Karyotype is presented by arranging the chromosomes of the somatic complement in a descending order of size keeping the longest chromosome in the extreme left and the shortest chromosome in the extreme right. In karyotype, usually the length of chromosomes, position of centromere, arm ratio, presence of secondary constriction and size of satellite bodies etc. are studied.

In rice somatic karyotype, small differences in absolute length, relative length and arm ratio of different chromosomes (Table 4.2) have been reported by several workers (Hu, 1964 ; Kurata and Omura, 1978; Kurata and Omura, 1982; Chen and Wu, 1982; Chung and Wu, 1987). The haploid chromosome complement (12) of *O. sativa* was classified into five large, four medium and three small chromosomes (Rau, 1929). Chromosome 10 is the only nucleolus organizer chromosome (*i.e.* having a nucleolus in the secondary constriction site of chromosome). Nucleolus is the active site for ribosomal RNA synthesis. Chromosome length of rice has been shown to range from 0.7 to 2.8 μm (Nandi, 1936) or 2.0 to 5.0 μm (Yasui, 1941).

Of the 12 chromosome pairs, 2 pairs were median (centromere at the centre), 9 pairs were submedian (centromere at one side of central point) and 1 pair subterminal (centromere located close to one end). The somatic chromosomes at prometaphase stage (*i.e.* before metaphase) are much larger (2.4 to 6.9 μm) than at metaphase stage (1.6 to 4.3 μm) (Sen, 1963). Kurata and Omura (1978, 1982) and Kurata *et al.* (1981a) used Giemsa-banding (G-banding)

Table 4.2: Mean values of absolute length (AL) (μm), relative length (RL) (μm) and arm ratios (AR) of somatic chromosomes in *O. sativa*

	Chromosome												Material	Stage	Reference
	1	2	3	4	5	6	7	8	9	10	11	12			
A L	4.3	3.7	3.3	2.9	2.9	2.6	2.6*	2.0*	2.8	2.3	2.3	1.8	Root tips	Metaphase	Hu (1964)
R L	13.0	11.1	9.8	8.7	8.6	7.8	7.8	6.0	8.3	6.7	6.7	5.4			
A R	2.2	1.5	1.3	1.1	1.7	2.1	2.1	3.0	2.5	1.6	1.3	2.1			
R L	15.2	12.2	11.2	9.6	7.6	7.6	7.0	6.5	6.2	6.2*	5.5	5.1	Root tips	Prometaphase	Kurata and
A R	2.0	1.7	1.4	3.5	2.2	1.3	1.2	1.4	1.6	2.6	1.3	2.0			Omura (1978)
R L	14.0	11.1	11.7	9.3	8.0	8.1	6.9	6.6	6.9	6.5*	5.5	5.3	Root tips	Prometaphase	Kurata and
A R	1.8	1.8	1.3	4.1	2.2	1.2	1.3	1.3	1.8	4.8	1.4	2.4			Omura (1982)
A L	6.9	5.9	5.2	4.4	4.2	3.8	3.4	3.2	3.0	2.7*	2.7	2.4	Root tips	Prometaphase	Chen and
R L	14.3	12.3	10.9	9.3	8.7	7.8	7.2	6.8	6.3	5.6	5.8	5.1			Wu (1982)
A R	1.9	1.5	1.9	4.6	1.5	1.7	1.6	1.4	2.2	3.7	1.2	2.6			
A L	6.4	5.8	5.4	4.3	4.2	3.8	3.6	3.1*	3.1	2.9*	2.7	2.5	Root tips	Prometaphase	Chung and
R L	13.8	11.7	11.1	9.2	8.7	7.9	7.3	6.7	6.5	6.0	5.8	5.3			Wu (1987)
A R	1.8	1.3	1.9	3.7	1.8	1.2	2.0	3.1	1.4	2.9	1.8	1.4			

*sat chromosomes

techniques to characterise somatic chromosomes of japonica rice variety of *O. sativa* at prometaphase stage. They observed that the length of chromosomes at metaphase ranged from 1.0 to 1.9 µm and were smaller than prometaphase chromosomes (1.5 to 4.1 µm).

The chromosomes were arranged in descending order of length and designated as K1 to K12. The karyotype consisted of five metacentric pairs, five submetacentric pairs and two subtelocentric pairs (Table 4.3). All chromosomes had deeply stained regions on both sides of the centromere indicating the presence of heterochromatin.

O. sativa and *O. glaberrima* have identical karyotypes. Kurata and Omura (1982) studied the somatic karyotypes of four species *viz. O. sativa, O. perennis, O. punctata* and *O. officinalis* and did not observe much difference among them. Unlike other species which have one pair of nucleolus organizer (no. 10), *O. perennis* possesses two pairs of nucleolus organizers (no. 8 and 10).

Pachytene Karyotype

Pachytene is the third substage of prophase I of meiosis. Pairing occurs among homologous chromosomes and bivalents (paired structure) are observed at this substage. Khush *et al.* (1984) described the major distinguishing features of each rice chromosome. In addition to the length, the centromere position, size, number and position of chromomeres were described (Table 4.4). Chromomeres are bead like structures found in chromosomes and their positions are constant. Chromatin is present in a highly condensed state in chromomeres. They stain darkly.

According to Khush *et al.* (1984), chromosome 9 is the nucleolar organizer but Chung and Wu (1987) designated chromosome 10 as the nucleous organizer. Some nucleolus organizing activity is also shown by chromosomes 1 and 11. Chromosome 10 is sometimes attached to the nucleolus. Short arms of chromosomes 4 and 11 stain darkly and are entirely heterochromatic. Chromosome 8 can be easily identified because of its dark-stained chromomeres.

Table 4.3 : Mean values of relative length (RL) (μm) and arm ratios (AR) of 12 somatic chromosomes in *Oryza sativa, O. perennis , O. punctata and O. officinalis.*

Chromosome	*O. sativa*		*O. perennis*		*O. punctata*		*O. officinalis*[a]	
	RL	AR	RL	AR	RL	AR	RL	AR
K1	14.0	1.82 sm[a]	13.1	1.90 sm	13.4	1.92 sm	13.2	1.76 sm
K2	11.1	1.80 sm	10.7	1.90 sm	11.0	1.91 sm	10.5	1.85 sm
K3	11.7	1.32 m	11.5	1.28 m	12.3	1.35 m	11.7	1.23 m
K4	9.3	4.08 st	8.8	3.64 st	9.3	3.54 st	9.2	3.48 st
K5	8.0	2.18 sm	8.3	1.93 sm	8.0	2.26 sm	7.8	1.91 sm
K6	8.1	1.17 m	8.8	1.22 m	8.0	1.17 m	8.8	1.22 m
K7	6.9	1.30 m	7.0	1.48 m	6.7	1.57 m	7.7	1.22 m
K8	6.6	1.31 m	6.7	1.33 m	6.4	1.17 m	6.8	1.21 m
K9	6.9	1.83 sm	7.1	1.80 sm	7.0	1.95 sm	7.2	1.89 sm
K10[b]	6.5	4.85 st	6.9	4.57 st	7.3	5.38 st	6.5	4.35 st
K11	5.3	1.35 m	5.6	1.36 m	5.4	1.39 m	5.3	1.24 m
K12	5.3	2.40 sm	5.6	1.93 sm	5.3	2.19 sm	5.4	1.75 sm

From Kurata and Omura (1982).

a Twenty prometaphase cells from *O. sativa,* 10 prometaphase from *O. perennis,* 5 prometaphase and 5 early metaphase from *O. punctata* and 4 prometaphase and 1 early metaphase cells from *O. officinalis* were measured.

b. Length of K10 was measured without secondary constriction and the satellite. m = median; sm = submedian and st = subterminal.

Considerable variations in absolute length and arm ratios of different chromosomes have been reported by various workers. The mean absolute length of the longest and the shortest chromosomes was 79.0 and 18.0 μm (Shastry *et al.* 1960). Discrepancies in the results for the absolute length and the arm ratio of same chromosome were observed because some investigators took the measurements at early pachytene stage when the chromosomes are longer, while others took the measurements at mid or late pachytene stage when the chromosomes are condensed and smaller in size.

Table 4.4: Diagnostic cytological features of pachytene chromosomes

Chromosome	Centromere position	Chromomeres						Remarks
		Centric		Median		Terminal		
		L	S	L	S	L	S	
1	Submedian	1	1		4			Occasional puff-like structure on long arm
2	Median	1	2	4	1			
3	Median	1	1	1	4	1	1	
4	Submedian	1	2	2	5			Short arm largely heterochromatic
5	Median	1	2		1			
6	Median	1	1	4	1			
7	Submedian	1	2	1	3	1		
8	Median	1	1	3	4	2		Chromomeres stain darkly
9	Subterminal	3	2	1	2		1	Nucleous organizing chromosome
10	Submedian	4	2	1			1	Sometimes attached to nucleolus
11	Median	1	1	2	2		4	Short arm largely heterochromatic
12	Submedian	2	1	4	3			

From Khush *et al.* (1984). L - Large; S - Small.

REFERENCES

Chang, T. T. 1964. Present knowledge of rice genetics and cytogenetics. *Technical Bulletin,* number 1, IRRI, Philippines, 96.

Chen, Y. and Wu, H.K. 1982. A comparison of Karyotypes among six *Oryza* species. *Bot. Bull.* Acad., Sinica, 23: 163 - 183.

Chung, M. and Wu, H. K. 1987. Karyotype analysis of IR36 and two trisomic lines of rice. *Bot. Bull.* Acd. Sinica, 28: 289 - 304.

Hu, C. H. 1964. Further studies on the chromosome morphology of *Oryza sativa* L. *Rice genetics and cytogenetics,* Elsevier Publishing, New York, 51-61.

Hu, C. H. 1970. Cytogenetic studies of *Oryza officinalis* complex III. The genomic constitution of *O. punctata* and *O. eichingeri*. *Cytologia*, 35: 304-318.

Katayama, T. 1976. Cytogenetical studies on the genus *Oryza* IX . The F1 hybrids from synthesized amphidiploid X species with BBCC genomes. *Japanese J. Genet*, 52: 301-307.

Katayama, T. and Ogawa, T. 1974. Cytogenetical studies on the genus *Oryza*. VII. Cytogenetical studies on F1 hybrids between diploid *O. punctata* and diploid species having C genome. *Japanese J. Breeding*, 24: 165-168.

Khush, G.S., Singh, R. J., Sur, S.C. and Librojo, A. L. 1984. Primary trisomics of rice: origin, morphology, cytology and use in linkage mapping. *Genetics*, 107: 141-163.

Khush, G.S. and Singh, R.J. 1991. Chromosome architecture and aneuploidy in rice. *Chromosome engineering in plants : genetics, breeding, evolution*, Part A. Elsevier Publishers B. V., Amsterdam, 578-598.

Kurata, N., Iwata, N. and Omura, T. 1981a. Karyotype analysis in rice. II. Identification of extra chromosome in trisomic plants and banding structure on some chromosomes. Japanese J. Genet., 56 : 41-50

Kurata , N. and Omura, T. 1982. Karyotype analysis in rice. III. Karyological comparisons among four *Oryza* species. *Japanese J. Breeding*, 32 : 253-258.

Kuwada, Y. 1910. *A cytological study of Oryza sativa* L. *Shokubutsugaku Zasshi*, 24 : 267-281.

Li, H. W., Weng, T.S., Chen, C.C. and Wang, W.H. 1961. Cytogenetical studies of *Oryza sativa* L. and its related species. *Bot. Bull.* Acad. Sinica, 2: 79-87.

Li, H.W., Weng, T.S., Chen, C.C. and Wang, W.H. 1962. Cytological studies of *Oryza sativa* L. and its related species. *Bot. Bull.* Acad. Sinica, 3: 209-219

Nandi, H.K. 1936. The chromosome morphology, secondary association and origin of cultivated rice. *J. Genet.*, 33: 315-336.

Nezu, M., Katayama, T.C. and Kihara, H. 1960. Genetic study of the genus *Oryza*. I. Crossability and chromosomal affinity among 17 species. *Seiken Ziho*, 11: 1-11.

Nowick, E.M. 1986. Chromosome pairing in *Oryza sativa* X *O. latifolia* Desv. hybrids. *Canadian J. Genet . Cytol.*, 28: 278-281.

Ogawa, T. and Katayama, T. 1973. Cytogenetical studies on the genus *Oryza*. VI. Chromosome pairing in the interspecific hybrids between *O. officinalis* and its related diploid species. *Japanese J. Genet.*, 48: 159-165.

Rau, N.S. 1929. Further contributions to the cytology of some crop plants of South India. *J. Indian Bot. Soc.*, 8 : 201 - 206.

Sen, S.K. 1963. Analysis of rice pachytene chromosomes. *The Nucleus*, 6 : 107-120.

Shastry, S.V.S., Ranga Rao, D.R. and Misra, R.N. 1960. Pachytene analysis in *Oryza*. I. Chromosome morphology in *Oryza sativa*. *Indian J. Genet.* 20 : 15-21.

Yasui, K. 1941. Diploid bud formation in a haploid *Oryza* with some remarks on the behaviour of nucleolus in mitosis. *Cytologia*, 11 : 515-525.

5

Rice Aneuploidy

Introduction

An organism having irregular multiple of the basic chromosome number(x) is called an aneuploid and the situation is called aneuploidy. The aneuploid usually possesses one or a few chromosomes more or less than the normal somatic chromosome number. There are different types of aneuploids—trisomic, monosomic, nullisomic etc.

A trisomic (2x+1) carries one chromosome more than the normal somatic chromosome complement whereas a monosomic (2x–1) contains one chromosome less than the somatic chromosome number. In the case of a nullisomic (2x–2), a pair of chromosomes is missing from the somatic chromosome number. A trisomic is known as a primary trisomic if the extra chromosome is the same as one of the haploid genome.

Of the various types of aneuploids, only primary trisomics have been well investigated in rice (Khush, 1975; Khush *et al.,* 1984). Trisomics can be produced spontaneously or from triploids (3x), or translocation heterozygotes or tetrasomics (2x+2) or from asynaptic and desynaptic plants. In rice like other crop plants, triploids have proved to be the best source of trisomics. The number of primary trisomics in a species is equal to the gametic chromosome number (n). In rice, the number of primary trisomics is 12.

Origin of Primary Trisomics

In rice most of the primary trisomics have been obtained from the triploid (3x) plants. Plants with varying number of extra chromosomes are produced by trisomics. Rice can tolerate a maximum of six extra chromosomes. A vast majority of the plants of triploid progenies are 2x (disomic), 2x + 1 (trisomic), 2x + 2 (tetrasomic), 2x+3 and 2x+4 plants (Khush and Singh, 1991).

Cytological Identification

The trisomics (2x+1) were first characterised into 12 groups on the basis of plant morphology (Khush and Singh, 1991). The extra chromosomes of the trisomics were identified on the basis of either somatic karyotype or pachytene karyotype. The trisomic having chromosome 1(longest chromosome) in triplicate (3 copies) is called triplo 1 and that having the extra chromosome 2 in triplicate is called triplo 2 and so on. A small nucleolus is associated with the extra chromosome in triplo 2 whereas in triplo 12, several small nucleoli in addition to the main nucleolus are observed (Khush *et al.*, 1984).

Morphological Identification

The trisomics (2x+1) differ from the disomics (2x) in many characters (Table 5.1). Trisomics for the long chromosomes (1 to 4) are more distinct from the disomics and from each other because the former are slower in growth (Khush and Singh, 1991). It is natural because the longer chromosomes cause more genetic imbalance than the shorter ones.

Some trisomics such as triplo 1 and triplo 8 can be distinguished at seedling stage (2-3 leaf stage). Triplo 1 seedlings have narrow, light green leaves with mottled appearance but triplo 8 seedlings have narrow, dark green, rolled leaves. Triplo 2, triplo 3, triplo 4, triplo 5, triplo 6, triplo 7 and triplo 9 can be distinguished from the disomics and from each other at 2-3 weeks after transplanting on the basis of their slower growth rate and distinct leaf features.

At seedling stage it is difficult to distinguish triplo 10, triplo 11 and triplo 12 because their growth rates are almost normal. At flowering stage, triplo 10 and triplo 12 can easily be distinguished. Triplo 10 has finer foliage and stem with narrow slender grains. Triplo 12 is about 20% taller than its disomic sibs and has a somewhat spreading habit. Triplo 11 is difficult to identify with certainty at any stage. Cytological examination is necessary to identify the triplo 11 plants. Sometimes, triplo 11 plants can be identified on the basis of their brownish glumes at the time of anthesis.

Seed Fertility

Trisomics for the longer chromosomes (triplo 1 to triplo 4) have very low seed fertility. Triplo 1 to triplo 3 give low seed set when open pollinated. But triplo 4 is completely self-sterile. The remaining trisomics produce abundant seed, although most of them have varying sterility (Khush and Singh, 1991).

Ligule Length

Ligule length can be used in distinguishing the different trisomics. Ligule is a thin, upright membrane present at the attachment site of the leaf sheath to the node of stem. Triplo 1, triplo 5 and triplo 8 have much reduced ligules whereas the ligules of triplo 3, triplo 6 and triplo 12 are longer than those of the diploid.

Grain Size and Awn

The grain size and shape of each trisomic are also specific. The grains of triplo 1 are slender and somewhat triangular and the glumes do not close properly. Triplo 2, triplo 4, triplo 8 and triplo 10 have short grains. The grains of triplo 6 are longer than those of the diploid. The grains of triplo 9 are the largest.

Table 5.1: Diagnostic morphological features of primary trisomics of rice

Primary trisomics	Growth habit	Leaf characters	Other characters
Triplo 1	Short, grassy	Long, narrow, thin leaves	Seed fertility low, grains narrow and triangular, late flowering
Triplo 2	Short, few tillers	Short, thick, dark green leaves	Short ligule, small anthers, short panicle, round grain, long glume, high female fertility but highly, self-sterile
Triplo 3	Short, many tillers	Short, thick,semi-rolled leaves	Awned, lax panicle, highly selfsterile, long ligule,early flowering
Triplo 4	Short, slow growth	Short, thick, dark green leaves	Late flowering highly sterile .
Triplo 5 leaves	Short	Twisted, short, fine white, shiny hairs on leaves at maximum tillering stage	Short ligule, compact and short panicle, short anthers, high self seed fertility
Triplo 6	Bushy, short, many tillers	Light green (pale) leaves at pre-flowering stage	Lax panicle, few grains, terminal grains deformed, long narrow grains, high self-fertility

Table 1. : *(Contd...)*

Primary trisomics	Growth habit	Leaf characters	Other characters
Triplo 7	Short, few tillers	Narrow, dark green, rolled leaves	Short ligule, incompletely exserted panicles, lax and partially self-fertile, long grains with tip awns
Triplo 8	Short,few tillers, slowest growth recovers at maximum tillering stage	Narrowest, dark green, rolled leaves	Short ligule and panicles, fertile, round grains
Triplo 9	Short, many tillers, spreading habit	Thick, short, dark green leaves	Panicle lax, large grains, highly self-fertile, panicles partially exserted
Triplo 10	Short, slender	Normal	Hairy glume, small and narrow grains, completely self-fertile
Triplo 11	Normal	Normal	Normal
Triplo 12	Spreading growth habit, tall, many tillers	Long,droopy, light green leaves	Panicles lax, grains of upper half of panicles possess short awn while lower ones have no awn

From Khush and Singh (1991).

Long awns are a constant feature of triplo 3. The grains of triplo 7 and triplo 12 have small awns.

Transmission Rate of Extra Chromosome

The transmission of extra chromosome from parents to progenies may occur through the female or the male parent. The longer chromosomes (1 to 4) do not transmit through the pollen (male) as they probably cause greater imbalance (Table 5.2). Transmission rates through female are lowest for triplo 1 and triplo 4 (15.5 and 17.8%). The transmission rate of extra chromosome through female parent ranges from 15.5 to 43.9 per cent.

The transmission through the male parent ranges from 0 to 27.3 %. Triplo 8 and triplo 9 give the highest transmission through the male.

Table 5.2: Female and male transmission rates of the extra chromosome in primary trisomics

	Transmission rates							
	(2x + 1) x 2x				2x x (2x + 1)			
Trisomic	Total	2x	2x+1	%2x+1	Total	2x	2x+1	%2x+1
Triplo 1	194	164	30	15.5	162	162	0	0.0
Triplo 2	364	247	114	31.6	20	20	0	0.0
Triplo 3	2414	1499	915	37.9	20	20	0	0.0
Triplo 4	326	268	59	17.8	0	0	0	0.0
Triplo 5	2067	1392	675	32.7	197	195	2	1.6
Triplo 6	1868	1176	692	37.1	209	198	11	5.6
Triplo 7	1404	967	437	31.1	113	113	0	0.0
Triplo 8	1755	1307	448	25.5	192	168	24	14.3
Triplo 9	1199	773	426	35.6	168	132	36	27.3
Triplo 10	1932	1402	530	27.4	209	206	3	1.5
Triplo 11	199	120	79	39.7	192	187	5	2.7
Triplo 12	1521	853	668	43.9	196	195	1	0.5

From Khush *et al.* (1984).

Table 5.3: Location of genes and isozyme loci through trisomic analysis on rice chromosomes

Genes and isozyme loci	Located on Chromosome Number
Genes	
eg and mp - 1	1
tri	2
cl, Ps-1, ws, mp-2, ch and bc-4	3
ch-1, dl and Hg	4
gh-1, gl-1, nl-1 and eui	5
spl-1	6
g	7
v-8	8
dp-2 and I-Bf	9
ygl, fl and pgl	10
la and z-2	11
lg and aul	12
Isozyme loci	
Got-1, Est-5 and lcd-1	1
Amp-1	2
Cat-1, Est-2, Pox-5, Amp-3 and Pgi-2	3
Pgi-1	4
Acp-1 and Sdh-1	6
Est -8	7
Amp-2 and Amp-4	8
Pgd-1 and Adh-1	11

Use of Trisomics in Locating Genes

Primary trisomics in rice have been successfully utilised for locating genes to specific chromosomes. The procedure consisted of crossing the plants with marker genes with all the primary trisomics and studying the segregation ratios in the F_2 or backcross progenies. Eleven of the twelve such progenies segregated in a regular diploid manner but ratios in one critical progeny were modified. In the critical trisomic progenies, the diploid portion of the F_2 population segregated in a ratio of 8:1 for the recessive gene instead of the normal 3:1 ratio, irrespective of the transmission rate. Similarly, the diploid portion of the backcross populations segregated in a ratio of 2:1 instead of 1:1 ratio. The segregation ratio for the whole populations (2x and 2x+1) of the F_2 progenies segregating in a trisomic fashion were modified to approximately 12.5:1 instead of 3:1 (Khush and Singh, 1991)

Several genes and isozyme loci were located to respective chromosomes of rice through trisomic analysis (Table 5.3) (Khush *et al.,* 1984; Nakagahra and Hayashi, 1976 ; Sano and Morishima, 1984; Sano and Barbier,1985; Wu *et al.*, 1988).

REFERENCES

Khush, G.S. 1975. Rice. *Handbook of Genetics,* ed. by R.C. King, Vol.2 Plenum Press, New York, 31-58.

Khush, G.S., Singh, R.J., Sur, R.C. and Librojo, A.L. 1984. Primary trisomics of rice: origin, morphology, cytology and use in linkage mapping. *Genetics,* 107:141-163.

Khush, G.S. and Singh, R.J. 1991. *Chromosome architecture and aneuploidy in rice. Chromosome engineering in plants : genetics, breeding, evolution,* part A. Elsevier Science Publishers B.V., Amsterdam, 578-598.

Nakagahra, M. and Hayashi, K. 1976. Detection of esterase loci of *Oryza sativa* L. *Japanese J.Breeding,* 26 (Suppl. II) : 114-115.

Sano, R and Morishima, H. 1984. Linkage relationships among 6 polymorphic isozyme loci in rice cultivars. *Japanese J. Breeding,* 34 (Suppl. I): 252-253.

Sano, R. and Barbier, P. 1985. Analysis of five isozyme genes and chromosomal location of Amp-1. *Rice Genet. Newsl,* 2:60-62.

Wu, K.S., Glaszmann, J.C. and Khush, G.S. 1988. Chromosomal location of ten isozyme loci in rice (*Oryza sativa* L.) through trisomic analysis. *Biochem. Genet,* 26: 303 -320.

6

Hybrid Rice Breeding

Introduction

Rice breeders all over the globe were surprised in 1977 when Chinese rice scientists first reported the development and cultivation of hybrid rice. Work on hybrid rice was started in 1964. Currently 60% of the total rice area in China has been reported to be under hybrid rice. The yield potential of commercial hybrid rice in China is about 5.5-7.0 t/ha as compared to 4-5 t/ha for conventional varieties. In 1992 hybrid rice occupied nearly 18 million hectares in China with average yield of 6.5 t/ha. These varieties have shown 20-30 per cent higher yield and possess wider adaptability than conventionally bred rice varieties.

The phenomenon of heterosis in rice was first reported by Jones in 1926 (Virmani *et al.,* 1981). Since then several rice breeders have suggested exploiting heterosis commercially by developing hybrid rice varieties. Several cytoplasmic male sterile lines were developed (Carnahan *et al.,* 1972 ; Athwal and Virmani, 1972), but a low natural outcrossing rate on the male sterile plants discouraged those workers from continuing research on hybrid rice.

The success of hybrid rice in China encouraged the International Rice Research Institute (IRRI) to revive research on hybrid breeding in 1979 to explore its problems and potentialities outside China. As a result the national rice improvement programmes of several countries such as India, Indonesia, Thailand, Bangladesh, Pakistan, Philippines and South Korea have shown interest in exploring the prospects and problems of hybrid breeding approach.

Approaches

There are four approaches for producing hybrid rice. These are :

(1) Three-line method or CMS system
(2) Two-line method or PGMS and TGMS system
(3) One-line method or apomixis system
(4) Chemically induced male sterility method

Three-line method

In this method, three lines are used *viz.* the cytoplasmic male sterile line (A line), the maintainer line (B line) and the restorer line (R line) (see Figure 6.1). These three lines of an indica variety were first developed by Yuan Long Ping in China in 1973.

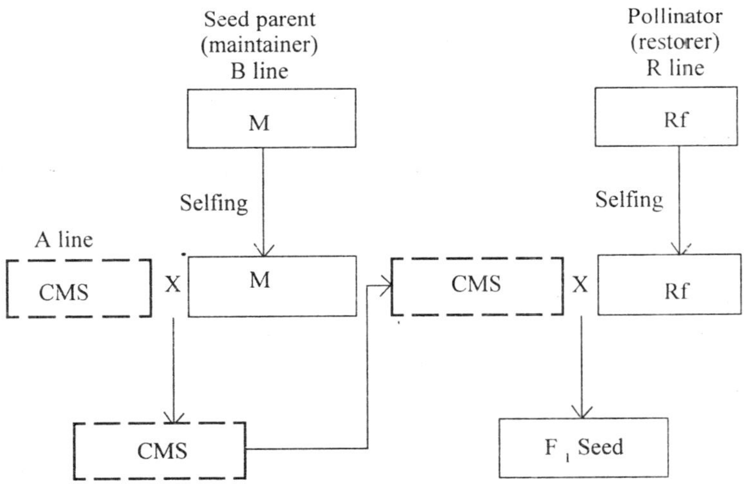

Fig. 6.1: Three-line method

Male sterility in the A line is determined by the cytoplasm, more particularly by gene(s) located in the mitochondria or plasmid like element. The A line is maintained by crossing it with the maintainer line or B line. The nuclear genotype of the B line is identical with the A line and the former possesses fertile cytoplasm unlike the A line. The R line is the pollinator parent having the fertility restorer gene (Rf) in the nucleus and is crossed with the A line to produce F_1 hybrid seed of rice. The A line is produced by repeated backcrossing of B line with the male sterile source used as female parent and the B line is used as recurrent parent (male parent) in each backcross. After BC_6 or BC_7 generations, the A line is produced having the nuclear genotype of B line and male sterile cytoplasm (see Figure 6.2).

Disadvantages of Three-line Method

There are several disadvantages of three-line method *e.g.*

(1) Development of the A line by conventional breeding method requires considerable time as it is developed by backcrossing B line six or seven times.
(2) Seed production of the A line is limited due to hybridization required between A line and B line.
(3) Small amount of hybrid seed from A x R cross is often produced due to low natural outcrossing.
(4) Male sterility of the A line is unstable and sometimes A line becomes male fertile resulting in the production of selfed seed on it instead of hybrid seed.

Two-line Method

Two-line method requires only two lines–male sterile line and restorer line to produce hybrid seed (Figure 6.3). No maintainer line is needed. The male sterile line possesses environmental genic male sterility (EGMS) system. EGMS is of two types :

(1) photo- period sensitive genic male sterility (PGMS)
(2) temperature sensitive genic male sterility (TGMS)

The PGMS line expresses male sterility only under certain day length *e.g.* the line 'Nangken 58S' shows male sterility when exposed to a day length more than 14 hours but becomes male

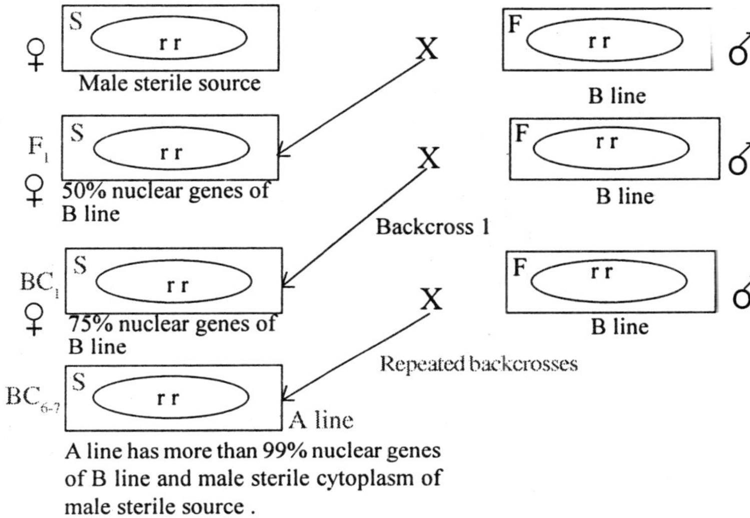

Fig 6.2: Production of A line

fertile at 13.75 hours day length at panicle initiation stage. However, evidence suggests that PGMS system is simultaneously regulated by photo-period and temperature.

In male sterile sources like 'Annong S-1' and 'H89-1', sterility is conditioned by TGMS system. Under high temperature (24-31°C), these lines show male sterility but revert to fertility under low temperature (18-24°C). TGMS was first discovered in a mutant line in Japan. This system has great potential for use in the tropics.

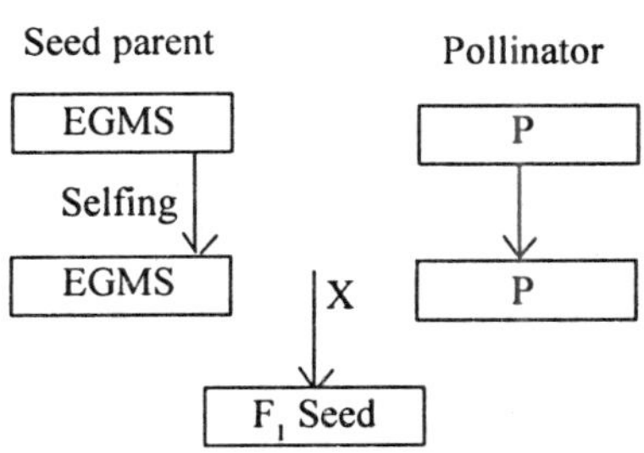

Fig. 6.3 : Two-line method

Advantages of Two-line Method

(1) The two-line method is easy and simple ; unlike three-line method this requires only male sterile line and restorer line to produce hybrid seed. Maintainer line is not required.
(2) Seed production of the male sterile line is not difficult because high temperature or specific photoperiod induces male sterility.
(3) Breeding cycle is rapid as this method does not involve backcrossing to develop male sterile line (A line) like three-line method.

Disadvantages of Two-line Method

(1) Seed parent expresses male sterility under specific temperature and photoperiod regime but reverts to fertility beyond the regime.
(2) TGMS system will not be effective in temperate regions.

One-line Method

One-line method is based on apomixis and somatic embryogenesis. Apomixis refers to the development of seed without fertilization. Apomixis is an asexual method of reproduction and it does not involve the fusion of male and female gametes. In apomixis the embryo develops from the mother plant. Apomixis renders some benefits in hybrid breeding. In the cultivation of hybrid varieties the farmers buy hybrid seeds every year and cannot retain seeds of previous crop due to segregation which results in very poor yield. Since the farmer purchases hybrid seed every year the cost of cultivation increases. This problem of procuring hybrid seed every year could be overcome by transferring the gene(s) governing apomixis to F_1 hybrid.

Advantages of One-line Method

(1) Once the F_1 hybrid is evolved, it could be multiplied

indefinitely without changing its heterozygous constitution through apomixis.

(2) The heterosis observed in the F_1 seed could be fixed and maintained apomictically without the risk of segregation in the subsequent generations. As a result the farmers can retain the progeny of F_1 hybrids as seeds for the next crop.

(3) The maternal effect could be permanently fixed through apomixis.

Apomixis in Relation to Hybrid Rice

In one-line method of hybrid rice production only the recurrent form of apomixis with obligate characteristic has great potential. Evidence suggests that apomixis is the result of an interaction of many genes and the process is unstable (Sharma, 1994). Apomixis, as a rule, is recessive to sexuality (amphimixis), although polyploid apomicts show tendency towards dominance (Stebbins, 1958).

Recurrent apomixis arises by any one of the following processes:

(1) **Diploid parthenogenesis :** The embryo develops from the unreduced egg cell without fertilization. Such egg-cell and all other cells in the embryo-sac are diploid (Figure 6.4).

(2) **Diploid apogamy :** The embryo develops from the diploid synergid or antipodal cell, so it is known as diploid apogamy.

(3) **Diploid apospory :** In apospory first diploid cell of ovule lying outside the embryosac develops into another embryosac without reduction. The embryo then originates from the embryosac that has developed from the diploid cell of nucellus or integument, this is called diploid apospory.

(4) **Adventive embryony :** The development of embryo directly from the diploid cells of ovule lying outside the embryosac belonging to either nucellus or integuments is referred to as adventive embryony. There is no production of another embryosac like diploid apospory.

Fig. 6.4 : Embryo-sac & other parts

Disadvantages of One-line Method

(1) Apomixis is unstable because the gene complexes conditioning apomixis often break down upon hybridization.

(2) Apomixis producing haploid embryo through haploid forms of parthenogenesis/apospory/apogamy is non-recurrent and has no utility in hybrid breeding. Such haploid embryos cannot perpetuate and are of academic interest only.

Chemically Induced Male Sterility Method

Male gametocides are the chemicals that can kill the male gametes (pollens). By using appropriate male gametocides any variety can be sterilized and used as a seed parent (female) in the hybridization programme. The chemicals that are found effective include : zinc methyl arsenate (4000 ppm), sodium methyl arsenate (2000 ppm), calcium sulphamate (2000 ppm) and fussol (fluoro acetamide at 2000 ppm). But ethrel was not found to be sufficiently effective.

Advantages

(1) The chemical method does not need the identification of a restorer line for making a commercial hybrid.

(2) There is no segregation for male sterility in the F_2 population, so the farmers can plant heterotic F_2 populations if needed.
(3) The male gametocides possess stable effect in causing male sterility.
(4) The chemicals have low residual toxicity in soil and on seed and are inexpensive.

Disadvantages

(1) The male gametocides are effective only at specific concentration, and have specificity with regard to time and method of application and stage of the plant.
(2) The effectiveness of the male gametocides is influenced by the weather and so chemical male sterility is not considered practical for large-scale seed production.

Somatic Embryogenesis in Producing F_1 Hybrid

Somatic embryogenesis refers to the process of development of somatic embryos. This process has great potential in the multiplication of F_1 seed like apomixis and is one of the ways for fixing heterosis permanently. Somatic embryos are the somatic cells with bipolar differentiation of root and shoot axis and are produced by *in vitro* techniques. A few embryogenic cells (somatic) present on the surface of the callus (undifferentiated mass of cells) undergo repeated mitotic division to produce somatic embryos. They are different from the zygotic embryos as the former does not result from the fusion of male and female gametes.

Once the F_1 hybrid rice is produced, it can be indefinitely multiplied through somatic embryogenesis. The somatic embryos can be easily encapsulated in a suitable gelling agent like sodium alginate. These encapsulated somatic embryos could be used as artificial F_1 seed of rice.

Major Characteristic of Hybrid Rice

The intervarietal hybrids involving indica and japonica sub-species developed experimentally in Japan, China and Korea have been reported to yield 10-15% higher than indica x indica hybrid and 20-25% higher than japonica x japonica hybrid.

The F_1 hybrid rice possesses some common characteristics:

(1) Hybrid rice, in general, yields nearly 20% higher than the best available conventional varieties.
(2) Hybrid rice has higher initial growth vigour.
(3) It has larger embryo which favours faster initial growth.
(4) Hybrid possesses increased total dry matter production potential.
(5) Harvest index of hybrid rice is high (ca. 0.6 and more)
(6) Hybrid also has better nitrogen absorption ability at tillering stage than the parents.
(7) The root system of the F_1 hybrid is stronger and more active and goes deeper into the soil.

Pre-requisites of a Successful Hybrid Programme

One of the most important pre-requisites for a successful hybrid rice programme is the effective and efficient production and distribution of F_1 seed of hybrid rice. For achieving this goal, field workers and seed growers must be trained by scientists. Proper isolation distance and planting ratio should be maintained in the seed production field.

Plot Layout in Three-line Method

Hybrid rice seed production using the cytoplasmic genetic male sterility system (three-line method) involves three stages *viz.* (Virmani *et al.*, 1981)

(1) multiplication of male sterile (A) and maintainer (B) lines,

(2) multiplication of restorer (R) line, and
(3) production of F_1 hybrid seed (A x R)

Two crossing plots are required—one for cms line production and the other for hybrid seed production.

In cms line production plot the maintainer B line (male) and the cms A line (female) are planted in the ratio 2:4 (see Figure 6.5). For a continuous supply of pollen grains, the B lines must be planted on two different seeding dates with considerable time interval.

In the case of hybrid seed production plot, the restorer R line (male) and the cms A line (female) are planted in the ratio 1:6. The R plants are equally planted on two seeding dates for a continuous supply of pollen grains (see Figure 6.6). This system is practised in China.

Techniques of Hybrid Seed Production

The hybrid seed production techniques as standardised in China and sucessfully implemented involve the following features :

(1) The seeding dates of the parents are adjusted to attain

Fig. 6.5. cms line production plot

□	□	X	X	X	X	O	O
□	□	X	X	X	X	O	O
O	O	X	X	X	X	□	□
O	O	X	X	X	X	□	□
□	□	X	X	X	X	O	O
□	□	X	X	X	X	O	O
O	O	X	X	X	X	□	□
O	O	X	X	X	X	□	□
B line (male)		A line (female)				B line (male)	

□ First seeding
O Second seeding

Fig. 6.6. Hybrid seed production plot

□	X	X	X	X	X	X	O
□	X	X	X	X	X	X	O
O	X	X	X	X	X	X	□
O	X	X	X	X	X	X	□
□	X	X	X	X	X	X	O
□	X	X	X	X	X	X	O
O	X	X	X	X	X	X	□
O	X	X	X	X	X	X	□
B line (male)	A line (female)						B line (male)

□ First seeding
O Second seeding

simultaneity of flowering.

(2) Male and female plants in the hybrid seed production plot are planted in the ratio 1 : 6.

(3) Planting is done across wind direction to facilitate pollination.

(4) Flag leaves of male sterile and restorer lines are clipped to expose their panicles for increased cross-pollination.

(5) Isolation distance of 40-100 m or isolation period of 21 days is maintained.

(6) Supplementary pollination using rope pulling method is done 3-5 times on a calm day during anthesis.

(7) GA_3 (20 ppm) is applied on the male sterile line and the restorer line at the initial heading stage to improve their panicle exertion.

Mixed Planting of cms and restorer lines

In order to increase the amount of F_1 hybrid seed, mixed seeding of cms and restorer lines can be done and this will increase the rate of outcrossing. However, the hybrid seed can be subsequently separated from the selfed seed of restorer line (pollinator parent) by either

(1) using seed sorting machine or

(2) incorporating a herbicide-sensitive gene in the restorer line (pollinator) and by spraying diluted herbicide after pollination to kill the pollen donor *i.e.* the restorer line.

Problems in Hybrid Rice Production

Rice is basically a self-pollinated crop due to its specific floral structure, anthesis and pattern of anther dehiscence. The extent of natural cross pollination varies from 0.1 to 4 per cent in cultivated varieties. However, the extent of outcrossing on male sterile plants is significantly high (15-75%). There are several constraints in hybrid rice production *e.g.*

(1) Sources of cytoplasmic male sterility for developing better cms lines are limited. Currently about 90% of the cms lines used in commercial indica hybrid production still belong to WA (wild abortive) source of cytoplasm whereas only BT cytoplasm is used in japonica hybrids. Lack of genetic diversity in the cms lines may lead the hybrid rice to a destructive disease or insect pest as was observed in hybrid maize production using Texas-cms source in the USA.

(2) Total number of elite restorer lines which can be used in commercial rice hybrids is also very limited. Lack of genetically diverse large number of elite restorer lines may make the rice hybrid susceptible to destructive diseases or pests.

(3) Heterosis level of japonica hybrids is not as high as indica hybrids and this is perhaps due to lower genetic diversity among japonica germplasm. As a result, the japonica hybrids have lower potentiality in increasing yield than the indica hybrids.

(4) Effective restorer lines in japonicas are not available.

(5) Lack of synchronization in flowering in the cms and restorer lines causes difficulties in hybrid seed production.

(6) Parents which are used to develop the hybrid rice varieties do not possess multiple disease and pest resistance. As a result the hybrids become susceptible to one or more diseases and/or pests.

(7) Grain quality of the existing hybrid rices is of medium level which is responsible for its low demand among consumers.

(8) Non-availability and poor performance of cms lines are major constraints in hybrid rice production in the developing countries.

(9) Adequate trained personnel are not available for producing hybrid rice.

(10) Infrastructure needed for hybrid rice production is inadequate.

(11) Cost of hybrid seed is very high. Socio-economic

conditions of the farmers in the developing countries do not permit them to afford the cost of hybrid seed and to apply high doses of fertilisers to hybrid rice for good yield.

(12) Lack of information on the nutrient uptake by the hybrid rice from different soil types may pose a serious problem. The farmers will get a good crop in the first year by growing the hybrid rice but the yield level will drastically reduce subsequently if the soil is not replenished with nutrients, in the form of manures and fertilisers, based on nutrient uptake study. Under subsistence agriculture, hybrid rice does not hold a good promise.

REFERENCES

Athwal, D.S. and Virmani, S.S.1972. Cytoplasmic male sterility and hybrid breeding in rice. *Rice Breeding*, IRRI, Philippines, 615-620.

Carnahan, H.L., Erickson, J.R., Tseng, S.T. and Rutger, J.N. 1972. Outlook for hybrid rice in USA. *Rice Breeding*, IRRI, Philippines, 603-660.

Jones, J.W. 1952. Improvement of rice. USDA year book of Agriculture, 415-441

Sharma, J.R. 1994. *Principles and Practice of Plant Breeding*. Tata McGraw-Hill Publishing Company Ltd., New Delhi.

Stebbins, G.L. 1958. Longevity, habitat and release of genetic variability in the higher plants. Cold Spring Harbour Symp. *Quant. Biol.*, 23 : 365-378.

Virmani, S.S., Chaudhary, R.C. and Khush, G.S. 1981. *Current Outlook on Hybrid Rice. Oryza*, 18 : 67 - 84.

7

Development of Hybrid Rice in India

Introduction

In the early seventies, the Chinese breeders identified the most stable source of cytoplasmic male sterility, what is known as "WA" (Wild Abortive) cytoplasm by a concerted effort, not by accidental discovery (Siddiq, 1993). Effective use of this male sterility source led to the development of some most productive hybrids in China. Nearly 58% of the total rice area in China was planted to hybrid rice in 1993.

Successful exploitation of hybrid rice technology by China aroused interests among Indian scientists. They explored the possibilities of directly introducing the Chinese hybrids, male sterile lines and restorer lines in India. All these attempts failed because all the male sterile lines became susceptible to different insect pests and diseases despite very poor plant stand in the field. Moreover, the same level of heterosis obtained in China could not be realised in India in a large number of hybrids. In addition to these defects, the Indian consumers did not prefer the coarse grain quality of the hybrids as the consumers are accustomed to fine grain rice.

The above experience made our scientists realise that development of male sterile lines in Indian varietal background is the foremost priority for exploiting hybrid rice technology in India.

Development of CMS lines

Various research institutes and universities in India and IRRI in the Philippines initiated research since early eighties to develop cytoplasmic male sterile (CMS) lines in the background of typical tropical rice varieties. As a result of these endeavours, as many as 22 CMS lines having the Chinese "WA" cytoplasm were developed in the background of tropical varieties. These CMS lines showed high stability and better combining ability in our conditions.

Development of CMS lines was possible because of the coordinated research efforts of 12 centres under National Research Network on "Development and Use of Hybrid Technology in Rice". The research network was coordinated by the Directorate of Rice Research at Hyderabad and it had 2 lead centres namely:

(1) PAU, Kapurthala in the Northern India.
(2) UAS, Mandya in the Southern India.

The lead centre at PAU, Kapurthala had four associate centres namely:

(1) HAU, Kaul
(2) Chinsurah
(3) NDUAT, Faizabad
(4) GBPUAT, Pantnagar

Similarly the lead centre at UAS, Mandya had three associate centres namely:

(1) TNAU, Coimbatore
(2) KKV, Karjat
(3) APPU, Maruteru

Under this network, there were three strategic research centres namely:

(1) DRR, Hyderabad : It also functioned as co-ordinating centre.
(2) CRRI, Cuttack
(3) IARI, New Delhi

A set of male sterile lines called "Punjab male sterile lines" was developed by the scientists of Punjab Agricultural University. The performance of these CMS lines varied from location to location. For instance, PMS 2A and PMS 3A did not perform well in Karnal while PMS 5A and PMS 8A showed poor performance at Mandya. At PAU, 10 male sterile lines in the series PMS 1A to PMS 10 A possessed "WA" source of male sterility.

In addition to these, many other CMS lines were developed in India involving "WA" source. These CMS lines included Madhu A, Bharati A, Pusa 33A, Pragati A, Hema A, Krishna A, Rajeswari A, Kiran A, Archana A, Madhuri A, Deepa A, Pusa 4-1-11A, Suphala A etc.

Two IRRI-bred CMS lines viz. IR62829A and IR58025A, both containing "WA" source, showed good performance in different test locations in India. At IRRI, the CMS line "IR66707A" was developed and it carried "ARC" male sterility source.

Different sources of male sterility were used in India, IRRI (Philippines) and China for the development of CMS lines. These sources include WA (Wild abortive), Rata, Kalinga, ARC and Dunghansali.

Development of Hybrids

Nearly 220 rice hybrids were developed in India by 1990 using the improved male sterile lines. This was one of the greatest achievements of the centres under National Research Network on hybrid rice. The hybrids belonged to different maturity groups varying from early variety to medium duration variety.

Most of these hybrids produced nearly 1 tonne more yield per hectare. The best performer hybrids varied from region to region. Even within the region, the same performer showed yield variation at different locations. In other words, the hybrids were location specific in performance which was not fully unexpected because of the large variation in each region due to geographical factors.

Hybrids with standard heterosis of 15-20% are generally rated the best. If this criterion is considered, the absolute yield gain in

Table 7.1: Best performed hybrids in the national rice trial

Hybrid	Days to 50% flowering	Yield (Kg/ha) Hybrid	Check	Yield increase (t/ha)
ANDHRA PRADESH				
Hyderabad				
IR 62829A/IR10198-66	90	6766	5258(Rasi)	1.51(28.7)
IR 62829A/IR 29723-143	90	6381	5258(Rasi)	1.12(21.3)
IR 62829A/ARC 11353	93	7328	6243(Rasi)	1.18(19.3)
IR 58025A/IR 29723-143	103	6188	5258(Rasi)	0.93(17.7)
Maruteru				
IR 62829A/MILYANG 46	75	4326	3397(IR 36)	0.91(27.3)
KARNATAKA				
Mandya				
IR 62829A/IR 9761-19	90	9502	8061(Annada)	1.44(17.9)
IR 58025A/IR 29723-143	90	7028	5744(Jaya)	1.28(22.3)
IR 58025A/IR 40750-82	103	6397	5923(Jaya)	1.10(20.8)
PUNJAB				
Kapurthala				
PMS 5A/PAU 1126-15-3	102	8545	7250(Jaya)	1.02(14.1)
PMS 3A/PAU 1106-6-2	106	8913	7259(Jaya)	1.39(19.2)
HARYANA				
Karnal				
IT 62829A/95		5580	4583(IR 36)	1.00(21.7)
IR 10198-66-2R				
PMS 10A/PAU 1106-6-2	104	10866	9233(LC)	1.60(17.7)

Note: Figures in parentheses are percent values.

Courtesy : Siddiq (1993)

Table 7.2: List of hybrids identified for on-farm verification trials

States	Hybrids
Andhra Pradesh	IR 62829A/Swarna, IR 62829A/Vajram,
	IR 58025A/Swarna, IR 58025A/Vajram,
	IR 58025A/IR 10198-66-2R, IR 62829A/IR 10198-66-2R
Punjab and	PMS 8A/PAU 1106-6-2, PMS 8A/IR 314332-6-2-1
Haryana	PMS 10A/PAU i106-6-2
Karnataka	IR 58025A/IR 9761-19-IR, IR 58025A/IR 29723-143-3R
Tamil Nadu	IR 58025A/IR 10198-66-2R, IR 62829A/IR 10198-66-2R
	IR 62829A/ARC 11353R
Uttar Pradesh	IR 62829A/Swarna, IR 62829A/WGL 3962

Courtesy: Siddiq (1993).

the productivity level would vary from region to region. As a result, it is better that a hybrid variety which gives one tonne more grain yield than the best commercial variety of similar duration is chosen for cultivation irrespective of the per cent yield heterosis in a particular region.

Research efforts have been directed within the country to improve the parental lines for producing highly heterotic hybrids and to diversify the cytoplasmic male sterility. Several hybrid combinations with better grain quality and improved resistance to major insect pests and diseases are on the brink of release in India.

Adoption of hybrid technology in our national research programme took unduly long time as compared to China, mainly because of the suspicion amongst the rice scientists regarding adaptability of this technology under our conditions. Many other factors also contributed to this delayed adoption.

Rice is a strictly self-pollinated crop with less than 0.1% outcrossing. It created some reservations among the scientists regarding percentage of hybrid seed set in male sterile line. This doubt is quite logical. But when a plant like rice is male sterile due to genetic or environmental factors, its stigma remains receptive for a longer period to receive pollen from adjacent plants. This increases the percentage of seed set. In China, the extent of outcrossing in male sterile plants was found as high as 40%. Under tropical climatic conditions like India, outcrossing was found to vary from 30 to 40%.

Research on Role of GA_3 in Hybrid Seed Production

Application of GA_3 (gibberellic acid) in the male sterile lines was recorded to increase seed set at IRRI and China. But GA_3 is the costliest input in hybrid seed production. Attempts are being made to partially substitute GA_3 in mixture with other less expensive chemicals like urea for increasing the hybrid seed production. The role of high wind velocity and favourable wind direction in seed set have not been properly understood despite a few isolated and stray reports. In the coming years, research must be directed towards these avenues.

Table 7.3: Relative efficacy of GA_3 and urea for increasing seed yield of CMS Line (IR 62829A)

Treatment	Yield (Kg/ha)					
	Hyderabad	Mandya	Faizabad	Pantnagar	Karnal	Delhi
GA_3 (60 ppm)	1638	1688	1244	1130	1772	1665
Urea (2%)	1295	1750	953	1180	1600	1557
GA_3 + Urea	1498	1756	1045	1373	1847	943
Most effective	1713	2475	1441	1506	2000	2056
Treatment	GA_3 + L	RP	GA_3 + L + RP	GA_3 + U + RP	GA_3 + U + L + RP	GA_3 + L + RP

Courtesy: Siddiq (1993).

GA_3 = GA_3 60 ppm at 10% flowering
U = Urea (2% spray)
L = Flag leaf cutting two days before panicle initiation
RP = Rope pulling twice a day at the time of anthesis.

Relative efficacy of GA_3 and urea for increasing the seed yield of the CMS line IR62829 A was studied at Hyderabad, Mandya, Faizabad, Pantnagar, Karnal and Delhi. GA_3 @ 60 ppm at 10 per cent flowering stage and urea spray @ 2 per cent were applied. Alongwith GA_3 and urea spray, effects of flag leaf cutting two days before panicle initiation and rope pulling twice a day at the time of anthesis were investigated with respect to seed set in CMS line. However, the results of different locations were not consistent.

Efforts towards Hybrid Basmati Rice

Scientists at the Indian Agricultural Research Institute developed "Pusa Basmati-1"—world's first high yielding dwarf basmati rice through strong endeavours of 20 years. This variety is widely cultivated in traditional basmati-growing areas.

Table 7.4: Yield range of hybrid seed during rabi and kharif 1990-91 in Hyderabad

Hybrid	Yield (Kg/ha)
RABI	
IR58025A/ PUSA 150R	1144
IR 62820A/CR 1359-8-IR	1732
IR 58025A/IR 35366-40-3-3-IR	2468
IR 62829A/IR 27315	1552
V 20A/IR 15324	837
V 20A/ARC 11353	314
KHARIF	
IR 62829A/IET 9844	1275
IR 62829A/Suweon 318R	964
IR 62829A/C 711097	690
IR 58025A/209- Chourigora	795
IR 58025A/HKR 119	1110
IR 58025A/IR 9761-19-R	545

Courtesy : Siddiq (1993).

Research has been underway to develop hybrid basmati rice. For this, attempts have been made to convert some potential basmati varieties into male sterile lines. The conversion has been easy because all the basmati types are maintainers. But the major hurdle lies in the fact that there is no restorer in the basmati germplasm. However, scientists at IRRI have identified several basmati derivatives having restoring ability.

In India, efforts are on to convert Pusa Basmati-1, Pusa-5615 and a few other basmati derivatives into CMS lines at Kapurthala (Punjab), IARI (Delhi) and Pantnagar (UP). Scientists are also trying to transfer restorer genes from non-basmati rice to basmati rice. The suspected restorer lines among the basmati derivatives are also being tested for using them as potential restorer lines in the production of hybrid basmati rice.

Indian scientists have also focussed their interests in the development of hybrids adapted to rainfed lowlands. But the prime concern in the recent years has been on the development of hybrids adapted to highly productive irrigated areas of north and north-western India and parts of the deltas in the south.

REFERENCE

Siddiq, E.A. 1993. Opportunities for hybrid rice in India. In : *Hybrid Rice : food Security in India.* Macmillan India Ltd., Madras, 28–43.

8

Global Status of Hybrid Rice Research

Introduction

Research work on hybrid rice was initiated by Prof. Yuan Longping in 1964. Since then much development has taken place in China and after 1980 outside China (Virmani, 1993). In this chapter I shall tell you about hybrid rice research in different countries except India. About the hybrid rice research in India, I shall discuss in a separate chapter.

By the year 2020 the world will shelter about 8 billion people, of which 40 per cent are estimated to be rice consumers (Chopra, 1994). In order to feed the exploding population, global annual rice production must increase by 74 per cent from the present 550 million tonnes to 900 million tonnes. Development of hybrid rice holds great promise to feed the global population.

Heterosis is the basis for the yield advantage of hybrid rice over conventional varieties. Among the available strategies, hybrid rice offers an immediate opportunity to break the yield plateau set by semi-dwarf rice varieties since mid 1960s, particularly in the irrigated ecosystems of the tropics. China has capitalised on hybrid rice technology extensively. Prospects outside China to exploit this technology look positive, especially in irrigated areas.

Hybrid rice technology exploits the phenomenon of heterosis or hybrid vigour which results in superior performance of F_1 generation of a cross between two genetically diverse parents. Hybrid rice was first commercially cultivated in China in the year

1976. Since then, China has increased hybrid rice area to 17.6 million hectares which accounts for ca. 55% of total rice area. Currently China is the only country in the world producing hybrid rice commercially which amounts to nearly 20% of the world's total rice production.

Initiation of Research

Since mid-sixties till recently, rice varietal improvement in different countries was primarily directed towards stabilising yield through build-up of disease and insect resistance, increasing per day productivity and improving grain quality in semi-dwarf rice varieties. The pattern was similar in India (Virmani, 1993). During 1970-72, attempts were made at IRRI to locate a source of cytoplasmic male sterility in indica rice for developing hybrid. But nobody at IRRI was convinced about the commercial feasibility of hybrid rice technology due to strict self pollinating nature of rice crop and low outcrossing potential, so the work was discontinued.

When China announced the successful development and commercialisation of hybrid rice technology, research on hybrid rice was revived at IRRI in 1979 to study the potential and problems of hybrid rice outside China.

Introduction of Hybrids to IRRI from China

Several hybrids from China were directly introduced to IRRI. But those hybrids did not perform well in the tropical conditions of IRRI due to their susceptibility to diseases and pests in comparison to IRRI's own best inbred varieties like IR 36 and IR 42. Attempts were also made to develop heterotic hybrids by crossing Chinese CMS lines and IRRI-bred restorers like IR 50 and IR 54 possessing multiple disease and pest resistance. Some of the heterotic combinations were identified at IRRI but those combinations failed to produce large amount of hybrid seed in a seed production plot. As a result, neither the rice hybrids nor the CMS lines introduced from China could be used as such to develop rice hybrids for tropical conditions (Virmani, 1993).

Development of CMS Lines at IRRI

Research was directed, after failure of Chinese hybrid and Chinese CMS lines, to develop CMS lines adapted to tropical conditions. Nearly 40 to 50 CMS lines were developed at IRRI of different genetic backgrounds and maturity groups. Of these, only two CMS lines namely IR 58025 A and IR 62829 A fulfilled the criteria of commercially usable CMS line and their seeds were shared with IRRI collaborators.

A commercially usable CMS line must possess at least four criteria *viz.*

(a) stable pollen sterility,
(b) adaptability to target environment (tropical or temperate),
(c) good combining ability, and
(d) good outcrossing potential.

In developing CMS lines at IRRI, breeders faced the problem of small number of maintenance lines for the available CMS system. Because of the few number of maintainer lines in the tropical rice cultivars, not many commercially usable CMS lines were developed at the beginning. So, IRRI started a specific maintainer breeding programme in 1990 which has now resulted in a large number of maintainer lines.

One of the IRRI-bred hybrids viz. IR64616H was tested at IRRI and in several collaborating countries. At no nitrogen level, the hybrid did not show yield difference with the best inbred cultivar but at high nitrogen level, the yield of hybrid was significantly higher than the inbred. Experimental results also revealed that the hybrid IR64616 H had disease and insect resistance as well as grain quality at par with the best inbred.

The hybrids IR64616H and IR64615H showed yield advantage over inbreds in India, Vietnam and Malaysia. A number of other hybrids, developed at IRRI, also showed yield superiority ranging from 9-21% in India.

Scientists at IRRI developed a heterotic rice hybrid derived from Chinese male sterile line, V20A and a Korean restorer variety,

Milyang 46 (an indica-japonica derivative). This hybrid outyielded the best inbred varieties by about 1 ton/ha but could not be commercialised in Korea because Korean consumers and farmers switched over their preference from indica-japonica variety to japonica variety. However, this hybrid variety V20A/Milyang 46 was released for commercial cultivation in China in the name of Wei You 46.

In Japan, rice scientists developed some heterotic rice hybrids with 10-15% yield advantage. These hybrid varieties were not released for commercial cultivation in Japan. The yield increase of the hybrids did not provide any additional profit because the market price of the high quality japonica inbred cultivars was higher than the yield advantage of the hybrids. In other words, the cost of per kilogram of hybrid rice with the 10-15% added yield was lower than the cost of high japonica cultivar. Japanese scientists have been trying to exploit indica/japonica heterosis using wide compatibility gene.

In the USA, hybrid rice research was conducted primarily by private companies in collaboration with China. Some Chinese hybrids were found to yield higher than the local check varieties but their grain quality did not find acceptance among the U.S. consumers. Moreover, the Chinese seed production system is labour intensive and it failed to become popular in the U.S.A. Research on hybrid rice in the U.S.A. could not gather much momentum over the years.

Male Sterility Used in Hybrid Rice Development

Most of the hybrid rice varieties developed in and outside China are based on cytoplasmic genetic male sterility system. This system requires a male sterile line (A), a maintainer line (B) and a restorer line (R). The use of this male sterility system is constrained due to several reasons:

(a) identification of maintainer and restorer lines,
(b) absence of restorer lines in japonica rices and
(c) few restorer lines in indica rices.

Japonica rice hybrids are produced by transferring restorer genes from indica rices to japonica rices. Later on, a search was geared-up to identify alternative system to develop rice hybrid. After considerable efforts, 2-line system was discovered.

Use of 2-Line System

The 2-line system is an alternative to the 3-line or cytoplasmic genetic male sterility system and has two modes namely TGMS (Thermosensitive genic male sterility system) and PGMS (Photoperiodsensitive genic male sterility system).

In TGMS, the line shows high to complete sterility at temperatures above 28°C. But at lower temperatures of 25-27°C, the line reverts to partial or full fertility. A temperature regime of 25-27°C prevails in the high altitude areas of the tropics and temperature above 28°C occurs in the plains of the tropics.

Thus a TGMS line offers 2 fold advantages:

(1) Seeds of the TGMS line can be multiplied in high altitude areas.

(2) Hybrid seeds can be commercially produced using TGMS line in the plains of tropical climate.

Norin PL12, a TGMS line developed by Dr. K. Maruyama and co-workers through mutation breeding in Japan, was brought to IRRI for developing hybrids. Results from Japan revealed no differences in the pollen fertility/sterility behaviour under different daylengths indicating that Norin PL12 is not photoperiod sensitive. But Norin PL12 showed normal fertility (93%) under a low temperature regime of 18-25°C and complete sterility in the high temperature regime of 24-31°C. So Norin PL12 mutant was designated as a thermosensitive genic male sterile line.

In a phytotron, Norin PL12 mutant can be grown to provide high temperature for sterility and low temperature for fertility. Since Norin PL12 from Japan was locked in a japonica genotype, attempts were made at IRRI to transfer TGMS gene to indica rices for use in tropics. IRRI scientists developed a TGMS mutant in indica rice

(IR 32364-20-1-3-2) which was then multiplied for providing seeds to different countries collaborating with IRRI.

In India, both TGMS and PGMS can be used to produce hybrid seeds because differences in temperature and daylength are found during rice growing seasons.

Exploitation of Tropical Japonica/Indica Heterosis

Scientists at IRRI have been endeavouring to raise rice yield by developing a new plant type in tropical japonica (javanica) background. The new plant type will possess 5-8 tillers but larger panicles with 200-300 spikelets and will be used for dense planting or direct seeding.

The new plant type tropical japonica rice will be used as one of the parents in crosses with semi-dwarf indica rice to develop rice hybrids with enhanced heterosis. Such breeding programme will necessitate the introduction of wide compatibility (wc) genes to make the cross between tropical japonica and indica rice a great success. Research work has already been initiated at IRRI to develop CMS and restorer lines and TGMS lines possessing wide compatibility genes. Success in the exploitation of tropical japonica/indica heterosis is likely to open a new chapter in rice breeding efforts to increase yield potential.

Limitations

Hybrid rice technology has not been fully exploited at a commercial scale in many countries except China. Several factors are responsible for the sluggish development of hybrid rice technology *e.g.*

(1) In many countries, development of stable male sterile line has taken more time than anticipated, mainly due to small number of maintainers among elite inbred lines. This constraint is now being overcome through a maintainer breeding programme specifically designed for the purpose.

(2) Many rice growing countries do not have a strong hybrid

seed industry. As a result, it is often difficult to produce the required quantity as well as quality of hybrid seed for distribution among farmers.

(3) Practice of 3-line system for hybrid seed production makes the seed very expensive and the farmers of developing countries cannot afford hybrid seed.

Future Research in Hybrid Rice

Rice scientists all over the globe should continue research in several avenues for developing highly heterotic rice hybrid in the coming years. Some of the potential areas are outlined below:

(1) Exploitation of tropical japonica/indica heterosis,
(2) Development of new and genetically diverse CMS lines, maintainer lines and restorer lines,
(3) Development of heterotic hybrids possessing at least 1 ton/ha yield advantage over the best inbreds,
(4) Improvement of outcrossing potential of male sterile lines through genetic manipulations, and
(5) Development of improved agronomic practices to realise the potential yield of hybrid rice and to reduce the seed rate.

Despite these research needs, a strong hybrid seed industry must be established in all rice growing countries. A close linkage between hybrid rice breeding programme and hybrid seed production agency at the public or private sector is a primary requisite for full exploitation of the outcomes of hybrid rice research.

REFERENCES

Chopra, V.L. and Narasimhulu, S.B. 1990. *Biotechnology for crop improvement. Genetic engineering and biotechnology—concepts, methods and applications,* Oxford & IBH Publishing Co. Pvt. Ltd., New Delhi, 159-180.

Chopra, K. 1994. Global assessment of hybrid rice technologies for breeding and seed production. Proceedings of the 18th session of International Rice Comm, 5-9 Sept, Rome, Italy.

Virmani, S.S. 1993. Global status of hybrid rice research and development. Hybrid Rice: Food Security in India, ed. by B.R. Barwale, Macmilan India Ltd., 3-22.

9

Rice Mutation Breeding in India

Introduction

Mutation breeding in rice was initiated by Ramiah and Parthasarathy (1938) in the year 1936 in India. Successful use of atomic energy in crop improvement by a large number of workers, particularly Swedish workers, aroused interests among the rice breeders all over the world (Chakrabarti, 1995). Subsequently a co-ordinated programme of mutation breeding in rice was initiated in south-east Asia in 1964 by International Atomic Energy Agency (IAEA).

Substantial work on mutation breeding in rice had been carried out in India during last decades in CRRI, IARI, BARC, Bose Research Institute and several other institutions.

Rice Varieties Developed through Induced Mutagenesis

In India, several rice varieties have been developed through induced mutations.These varieties are shown in Table 9.1.

Among the directly induced mutant varieties only the variety "Jagannath" released in 1970 became popular in low lying delta areas as a long duration variety. It was an induced semi-dwarf isolated from a popular long duration variety T-141 of Orissa. The variety "Jagannath" was not only directly released as a mutant variety but also used as a parent in hybridization with local tall and improved dwarf varieties. The variety "Keshari" released in 1980 was developed from the cross Kumar/Jagannath.

Table9.1: Mutant varieties developed in India

Mutant and hybrid mutant variety	Year of release	Parentage	Salient features
K 84	1968	T.65	Indica grained mutant
Jagannath	1970	T.141	High yielding, semitall & fertilizer responsive
IIT 48	1972	IR-8	One week earlier than IR-8
IIT 60	1972	IR-8	One month earlier than blast and cold resistant
Hybrid Mutant 95	1973	(Jhona 349 x TN1) Irradiated	Early, dwarf & high yielding
Sattari	1983	(NSJ 200 x Padma) Irradiated	Extra early
Keshari	1980	(Kumar x Jagannath)	Early maturity, medium slender grain

Courtesy : Chakrabarti (1995).

Development of PNR Rice Varieties

In 70s and 80s a series of PNR rice varieties were developed at IARI, New Delhi with the uses of induced rice mutants as parents in hybridization programme. Out of several PNR varieties, two varieties namely PNR 162 and PNR 381 were very promising and were released and notified throughout India for general cultivation. Gamma radiation was used as a mutagen to produce a large number of desirable induced mutants. The mutants were screened and the desirable ones were used as donors in the combination breeding programmes which led to the development of PNR rice varieties. Salient features of PNR rice varieties are given in Table 9.2.

Table 9.2: Salient Features of PNR rice varieties

Variety	Parentage	Maturity (days)	Height group	Yield (t/ha)
PNR-166	IR-8 Mutant x Basmati-370	85-110	Semidwarf	7.0
PNR-381	Tainan-3 Mut. x Basmati-370	85-105	Semidwarf	8.7
PNR-571	MW 10 Mut. x PNR-351	85-110	Semidwarf	5.8
PNR-162	Jaya Mut. x Basmati-370	80-115	Semidwarf	8.0
PNR-519	Tainan-3 Mut./ Basmati-370// PNR 417-3	90-120	Semidwarf	9.0
PNR-520	PNR-351 x PNR-140	115-125	Semidwarf	8.6
PNR-546	PNR 125-2 x PNR 130-2	90-120	Semidwarf	6.0
PNR-555-5 (J.D.3)	Dular Mut./ N-22//PNR 351/ Mtu 17 Mut.	65-70	Medium tall	5.3
PNR-555-28 (J.D.10)	-do-	80-85	Semidwarf	6.8
PNR-550-1-2 (J.D.8)	Dular Mut. x N-22 Mut.	70-75	Medium tall	4.2

Characteristics of PNR Varieties

PNR rice varieties have some characteristics which are described below :

(1) PNR rice varieties are short durational. Varieties like PNR-555-5 (J.D.3) and PNR-550-1-2-(or J.D.8) mature in 65-75 days. Other varieties like PNR-166, PNR-381, PNR-571, PNR-162 and PNR-519 mature in 85-120 days. J.D. is the abbreviated form of "Jaldi Dhan" (Extra early rice).

(2) PNR varieties are high yielders with average yield varying from 4.2 tonnes/ha in J.D.8 variety to 9.0 tonnes/ha in the case of variety PNR-519.

(3) These rice varieties are semi-dwarf except J.D.3 and J.D.8 which are medium tall.

(4) Semi dwarf varieties like PNR-381, PNR-519 and PNR-571 have been developed from crosses involving an alternative dwarfing source other than DGWG (Dee-geo-woo-gen) source. These varieties have great potential in widening the genetic diversity in the rice cultivars.

Development of Jaldi Dhan Varieties

Jaldi Dhan (J.D) varieties, a group of PNR varieties, are extra early in maturity (65-85 days). Mutations were induced in the popular traditional varieties like Nagina-22, Mtu.-17, Dular, Brown Gora and also in the newly improved varieties like MW10, Kalinga III to produce desirable mutants. Use of these mutants as parents in combination breeding resulted in a number of promising Jaldi Dhan (J.D) varieties. Jaldi Dhan varieties 1, 3, 5, 6 and 8 were found promising in eastern India.

Characteristics of Jaldi Dhan Varieties

Jaldi Dhan varieties have the following characteristics:

(1) The Jaldi Dhan varieties have high harvest index and high

productivity efficiency associated with extra early maturity.

(2) These varieties are widely adaptable to various rainfed upland areas (particularly drought-prone areas) and low input situations. So these varieties became very popular with the farmers of the tribal areas of Madhya Pradesh, drought and famine-prone Kalahandi district of Orissa or in the areas with problematic soils. The fruits of "Green revolution" could not reach these areas.

Limitations of Mutation Breeding

Mutation breeding could not project itself as an effective method like conventional breeding despite tall expectations made at the inception of mutation research for crop improvement throughout the world. Moreover, only a small fraction of induced mutants has been found suitable for direct release as varieties for commercial cultivation.

A mutant carrying several mutations showing positive changes is unlikely. So the promising use of mutants, instead of direct release as a variety lies in their use in hybridization programmes as parents to develop a new variety or to enhance the genetic diversity. Enhancement of genetic diversity is an urgent need of the time.

Prospects of Mutation Breeding

Despite the limited use of induced mutants either directly as a variety or as a parent in hybridization programme in India, the mutation breeding work could provide several benefits in the coming years.

Extensive use of DGWG (Dee-Geo-Woo-Gen) dwarfing gene from a spontaneous dwarf mutant from the rice variety "Woo-Gen" increased the yield potential of tropical rice in the era of green revolution and changed the agricultural scenario of India from grain importing status to grain exporting status. All these varieties and over 95 per cent of our present day cultivated semidwarf rice varieties possess the same dwarfing gene source directly or indirectly. This has led to genetic homogeneity among rice varieties and narrowed down genetic diversity, paving the way to ecological hazards and vulnerability to disease-pest. The problems are aggravated under high input-farming and humid monsoon climates. The urgent need of the present time is to seek alternative dwarfing sources from induced mutation breeding. In countries like USA and Thailand, uses of new source of dwarfness other than DGWG gene are being made to develop short statured high yielding rice varieties. The approach has great potential in increasing the genetic diversity in rice.

The potential of induced mutants to contribute to improved disease resistance in rice crop is well established (Yamasaki and Kawai, 1968; Swaminathan *et al.*,1971). For example, the rice variety PNR 381 with multiple resistance to pests and diseases was developed from a cross between Tainan-3 induced mutant and indica variety Basmati-370 at IARI. Mutation breeding holds great potential in the coming decades in the continuous search for new sources of resistance to diseases and pests in rice.

REFERENCES

Chakrabarti , S.N. 1995. Mutation breeding in India with particular reference to PNR rice varieties. *J. Nuclear Agric. Biol.,* 24(2) ; 73-82.

Ramiah, K. and Parthasarathy, N. 1938. Proceedings of 25th Indian Science Congress Abst., 12.

Swaminathan, M.S., Siddiq , E. A. , Palanichamy , K. , Vilawan , S. , Chakrabarti, S.N. , Puri, R.P. Singh, V.P. and Vairawan, S. 1971. Proceedings FAO/ IAEA Res. Coord. Meeting on "Use of induced mutations for rice improvement," UPCA, Los Banos, Philippines , 13.

Yamasaki, Y. and Kawai, T, 1968. Rice breeding with induced mutations, IAEA Tech. Rep. Ser. No. 86, 65.

10

Wide Hybridization in Rice

Introduction

Wide hybridization refers to hybridization between individuals from different species, belonging to the same genus or to different genera. It includes intergeneric and interspecific crosses.

From the early 1970s rice productivity has almost reached a plateau. Despite all breeding efforts no tangible improvement in rice yield has been obtained anywhere in the world. So transfer of useful genes from wild species of rice to cultivated rice appears to have great potentiality in getting a further jump in rice productivity.

Wild species of rice are a rich source of useful genes for the genetic improvement of cultivated rice. However, efforts to transfer useful genes from wild species to cultivated rice have been very limited. The only examples are the transfer of a gene for grassy stunt virus resistance from *O. nivara* (Khush *et al.*, 1977) and of a cytoplasm of *O. perennis* to produce cyto-sterile lines for hybrid seed production (Lin and Yuan, 1978). In recent years, disease and insect outbreaks on rice have become more intense and germplasm resources of rice, including wild species, have been evaluated to identify donors for resistance (Khush, 1984). Several wild species are highly resistant to all the known biotypes of brown plant hopper (BPH), white-backed plant hopper (WBPH), and green leaf hopper (Heinrichs *et al.*, 1985).

Oryza sativa and its closely related wild species *O. nivara, O. perennis* and *O. longistaminata* share the AA genome. The wild species can easily be crossed with *O. sativa* and genes from them can be transferred to cultivated rice by conventional crossing and backcrossing procedures. However, wild species with other genomes are difficult to cross and they produce completely male-sterile hybrids (Jena and Khush, 1990). Several workers have investigated the hybrids of *O. sativa* with *O. officinalis* which has the CC genome. Morinaga and Kuriyama (1958) and Li *et al.* (1964) observed no bivalent formation in the AC hybrids. These studies investigated genome homologies and species relationships, not the transfer of useful traits from *O. officinalis* to cultivated rice.

New Approaches :Transfer of High Biomass

Increase in rice yield may be possible by transferring genes for high biomass from wild species to cultivated species. High biomass will proportionately increase the harvest index, which is the ratio of economic yield to biological yield excluding the roots. It is well known that rice yield is positively correlated to harvest index.

Physiological characteristics concerned with yield are related to source (amount of carbohydrate or dry matter produced), sink size (number and size of florets) and harvest index. Under favourable conditions, sink size is a major limiting factor to increased yield, particularly in medium and long duration rice varieties because many non-structural carbohydrates (starch) remain in the stem and leaf sheaths at maturity and do not translocate to panicles. However, in an environment with full sunlight during the grain filling period, rice varieties with a longer growth duration can probably support larger panicles (sink) than those with a shorter growth duration if the translocation of pre-heading stored carbohydrates is efficient. In addition, panicle weight can probably be increased more satisfactorily by increasing grain number rather than grain size, although both are possible (Tran and Peterson, 1983). Increase in dry matter (source) results primarily from increased leaf area.

Transfer of genes for high biomass and leaf area from wild species to *Oryza sativa* (having AA genome) could be either (1) from the same AA genome of wild species *O. nivara* and *O.*

rufipogon or (ii) from the CCDD genomic constitution of the wild species *O. latifolia*. In the former approach, gene transfer should result from interspecific chromosome pairing of two AA genomes and recombinations in the F_1 and later generations. But in the latter approach, addition line production and the induction of translocations are used to facilitate transfer of genes influencing yield potential between different genomes.

Transfer of Drought and Salinity Tolerance

From the early 1970s, distant hybridization has been carried out in China between *O. sativa* (japonica or keng type) and *Sorghum vulgare* (Shanbao and Xiaolan, 1989) to improve drought and salinity tolerance in cultivated rice. The hybrids are stabilised after the tenth generation. Backcrossing has been combined with anther culture to speed up stabilization and selection. China has released for cultivation these stabilised hybrids which retain several sorghum characteristics such as better tolerance to drought, saline-alkali soils and blast. Possibly one or part of the sorghum chromosomes, not the whole genome, are incorporated in these stabilised hybrids.

The line A7927, which resulted from the cross A 7505 (sorghum) x Jingyin 83 (a keng rice variety), has a well-developed root system and is tolerant to pH = 8.5 and 0.3 percent or more salt content during the seedling stage. Its growth recovers quickly after irrigation and its 1000-seed weight is 30g. The line 837003 from the cross sorghum D 7846 x rice Nan 81 is tolerant to drought and resistant to lodging and disease. Its leaf blades roll inside in response to drought.

In order to develop salt tolerance in rice, efforts have been made in Colorado State University (USA) to develop a rice variety through protoplast fusion between rice and *Leptochloa fusca*. In Japan, salt tolerant rice plant has been regenerated from the hybrid callus resulted from the protoplast fusion of rice and *Echinochloa oryzicola*.

Transfer of Disease and Insect Resistance

Five diseases (blast, bacterial blight, tungro, sheath blight, grassy stunt) and four insects (brown plant hopper, stem borers, gall midge and green leaf hopper) are a common occurrence in most countries in tropical and subtropical Asia.

Breeding programmes should be undertaken to transfer useful resistant genes from wild species to cultivated species, *e.g.* bacterial blight resistant genes of *O. longistaminata* and *O. officinalis* may be transferred to cultivated rice. The wild species *O. australiensis* could be used as a donor for brown plant hopper resistant genes to cultivated rice. Crosses between *O. sativa* and *O. officinalis* were made to explore the possibility of transferring resistance to BPH from *O. officinalis* to *O. sativa* (Khush and Singh, 1991). For transferring useful genes from wild species to cultivated rice, monosomic alien addition lines (MAALS) would play a key role.

Problems of Wide Hybridization

Several problems have been encountered in wide hybridization in rice *viz.*

(1) cross incompatibility,
(2) endosperm abortion,
(3) hybrid sterility, and
(4) lack of meiotic recombination

Overcoming Problems of Wide Hybridization

(1) **Somatic Hybridization** : To overcome cross incompatibility between two species or genera, somatic hybridization can be accomplished by fusing their protoplasts. Protoplasts are obtained by enzymatic (cellulase and pectinase) digestion of cell wall. Protoplasts so obtained are fused in presence of polyethylene glycol (PEG) which results in a protoplast population consisting of a heterogeneous mixture of hybrid protoplasts

(heterokaryons) and unfused protoplasts (homokaryons). Hybrid protoplasts constitute only a small proportion of the mixture. The hybrid protoplasts are selected from the mixture by using specific selection schemes, such as the use of visual or biochemical selectable markers and are regenerated in *in vitro* conditions to produce plants called somatic hybrids.

(2) **Embryo Rescue** : Embryo rescue technique is followed when the wide hybridization fails due to endosperm abortion. To overcome the problem of endosperm abortion, 14-day old hybrid embryo is detached from the plant and cultured on a suitable culture medium. This helps the F_1 hybrid embryo to survive and it is regenerated into a whole rice plant.

(3) **Allotriploids** : Allotriploid hybrids having two genomes of the cultivated rice species and one genome of the wild species have great potential in overcoming hybrid sterility. The sterile hybrid does not flower and hence produces no seed. The allotriploids could be produced in rice by two methods (Jena and Khush, 1990).

(i) Occasionally the F_1 produces unreduced (2n) gametes and these 2n gametes are fertilised by an n gamete (Fig 10.1). This results in the production of allotriploid. This method is preferred because in the diploid hybrid there is more likelihood of pairing and recombination between the parental genomes.

(ii) In the second method, the autotetraploid of the cultivated species is hybridised with the diploid wild species (Fig. 10.2). This results in the production of triploids. In this method, there is preferential pairing between the homologous chromosomes of two identical genomes (two As) and the chances for homoeologous pairing and recombination are minimised.

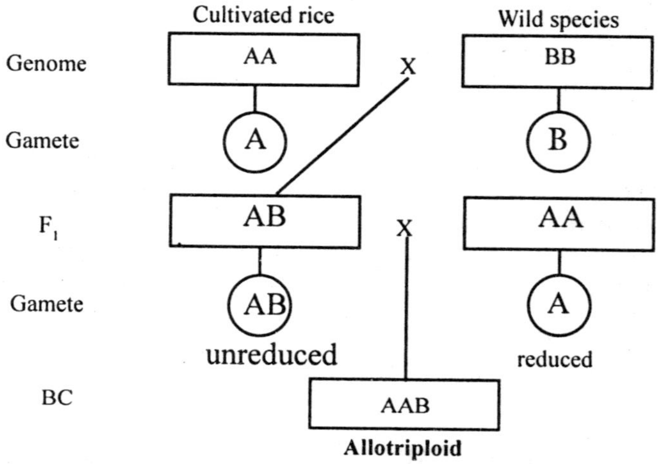

Fig. 10.1: Production of allotriploid (Method 1)

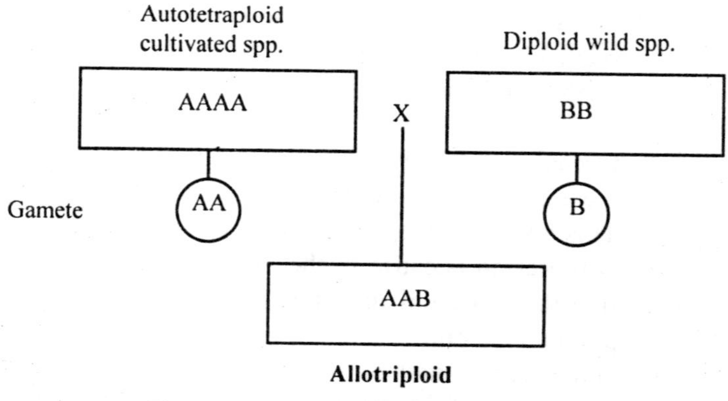

Fig. 2 Production of allotriploid (Method 2)

(4) **Tissue culture technique :** When there is little or no meiotic recombination, exchanges between the chromosomes of the cultivated rice and those of the wild species need to be stimulated. In such cases the hybrid may be passed through a tissue culture cycle which may cause chromosome breakage and rearrangements (Toenniessen, 1990).

Desirable Traits in Different Species

The primary goal of wide hybridization is the transfer of a small chromosome segment from the wild species into breeding lines. The wild species must possess one or more desirable genes instilling a new useful trait. There are 22 *Oryza* species, only two of which *viz. O. sativa* and *O. glaberrima* are cultivated. The genus contains six known genomes, A-F, combined into tetraploid species as well as diploids. The wild *Oryza* species have many genetic traits that would be of value in cultivated rice including resistance to diseases and pests and abiotic stresses (see Table 10.1).

Use of Molecular Probes in Wide Hybrid

Molecular probes refer to small segments of DNA or RNA of known base sequence. Molecular probes made from repetitive DNA sequences can be used to test for the presence of alien chromosomes or chromosome segments in the wide hybrid. The ideal probe is one whose DNA sequence is highly repeated and dispersed throughout the genome of the wild rice species but absent from that of cultivated rice (Toenniessen, 1990). For example, there are thousands of copies of the ribosomal genes in rice and different *Oryza* species have different sized spacer DNA sequences in the tandem repeats of these genes (Cordesse *et al*., 1990). Thus spacer sequences can be used as species- specific probes and for studies of evolution. Zhao *et al.* (1989) identified repetitive DNA sequences specific to the AA, CC, EE and FF genomes which can be used similarly. Ongoing research would soon provide DNA probes which can be used to follow the transfer of alien chromatin from any wild species used in a wide cross with cultivated rice. Using *in situ* hybridization of these probes and

> **Spacer DNA Sequence** : It is the untranscribed segment of eukaryotic genome flanking two functional genes (cistrons). Spacer sequence contains repetitive DNA and is important for chromosomal synapsis (pairing) during meiosis.

scanning electron microscopy it is also possible to visualise the alien chromatin on the rice chromosomes (Aswidinnour *et al.*, 1989).

Table 10.1. Desirable genetic traits identified in the wild relatives of *Oryza sativa*

Species	2n	Genome	Traits
O. nivara	24	AA	Grassy stunt virus resistance
O. rufipogon	24	AA	Cytoplasmic male sterility, tolerance to stagnant flooding
O. glaberrima	24	AA	GLH resistance, early vegetative vigour
O. barthii	24	AA	Bacterial blight (BB) resistance
O. longistaminata	24	AA	Floral characteristics for outcrossing
O. punctata	24, 48	BB, BBCC	BPH, WBPH, GLH resistance
O. officinalis	24	CC	BPH, WBPH, GLH resistance
O. eichingeri	24	CC	BPH, WBPH, GLH resistance
O. minuta	48	BBCC	BPH, WBPH, GLH, blast and BB resistance
O. australiensis	24	EE	BPH resistance, drought tolerance
O. brachyantha	24	FF	Rice whorl maggot and stem borer resistance
O. ridleyi	48	—	Rice whorl maggot resistance

Courtesy : Toenniessen, 1990. BPH—Brown plant hopper, WBPH—white backed plant hopper, GLH—Green leaf hopper.

Achievements

One of the major achievements of wide hybridization to date is the transfer of a gene for resistance to grassy stunt virus from *Oryza nivara*, a wild rice from India, to IRRI breeding lines (Toenniessen, 1990). It is providing good resistance in numerous lines cultivated throughout Asia and has saved farmers hundreds of millions of dollars in crop loss and pesticide costs. Advanced generation progenies with genes for resistance to brown plant hopper and white backed plant hopper from *Oryza officinalis* have been provided by IRRI to national programmes for field testing. Hybrids have been made at IRRI with *Oryza minuta* with a hope of transferring genes for resistance to blast and bacterial blight. Similar hybrids have been produced and are being tested in Korea and China. The male sterility gene used in China for hybrid rice seed production came from a wild relative. Hybrid seeds are now planted on over 10 million hectares and give ca. 15 per cent increased yield.

REFERENCES

Aswidinnour, H., Dallas, J.F., Dille, J.E., Gustafson, J.P., McIntyre, C.L. and Sears, E.R. 1989. Molecular determination of species-specific repetitive DNA sequences in rice . Third Annual Meeting of Rockefeller Foundation's International Program on Rice Biotechnology.

Cordesse, F., Second, G and Delseny, M. 1990. Ribosomal gene spacer length variability in cultivated and wild rice species. *Theor. Appl. Genet.*, 79 : 81-88.

Heinrichs , E. A. , Medrano , F. G. and Rapusas , H. R. 1985. Genetic evalutation for insect resistance in rice. IRRI, Philippines

Jena, K.K. and Khush, G.S. 1990. Introgression of genes from *Oryza officinalis* Well ex Watt to cultivated rice , *O. sativa* L. *Theor. Appl. Genet.*, 80: 737-745.

Khush, G.S., Ling, K.C., Aquino, R.C. and Aguiero, V.M. 1977. Breeding for resistance to grassy stunt in rice. Proceedings 3rd International Congress SABRAO, Canberra , Australia, *Plant Breeding Papers* 1(4) : 3-9.

Li, H.W., Chen, C.C., Lu, Katherine, C.L., Wu, H.K. and Hu, C.H. 1964. Pachytene studies of the hybrid *Oryza sativa* x *O. officinalis. Rice Genetics and Cytogenetics.* Elsevier Science Publishers, Amsterdam, 141-142.

Lin, S.C. and Yuan, L.P. 1978. *Hybrid rice breeding in China. Innovative approaches to rice improvement.* IRRI, Philippines, 35-51.

Morinaga, T. and Kuriyama, H. 1958. On the interspecific hybrid of *Oryza sativa* and *O. officinalis. Japanese J. Breeding,* 8:189.

Shanbao, C. and Xiaolan, D. 1989. Genetic variation in stabilised rice-sorghum crosses cultivated in a droughtprone area of China. International Rice Comm. Newsl., 38.

Toenniessen , G. H. 1990. Rice biotechnology : progress and prospects. SCI conference on pest management in rice, June 4-7.

Tran, D.V. and Peterson, M.L. 1983. Performance of near isogenic lines of rice differing in growth duration ; carbohydrate partitioning during grain filling. *Crop Science,* 23 : 243-246.

Zhao, X., Wu, T., Xie, Y. and Wu, R. 1989. Genome specific repetitive sequences in the genus *Oryza* . *Theor. Appl. Genet.*, 78 : 201-209.

11

Rice Ideotype Breeding

Introduction

An ideotype is a biological model that performs or behaves in a predictable manner in a defined environment. A crop ideotype is a plant model which is expected to yield a greater quantity or quality of grains, fibre, oil or other useful product when developed as a cultivar. The term ideotype was proposed by Donald in 1968 working on wheat.

Ideotype for Higher Yield

Since the evolution of dwarf rice varieties with increased yield, no significant achievement has been made by rice breeders worldwide over last three decades in breaking the yield plateau of rice. Rice scientists have been engaged in designing a rice ideotype for achieving a quantum jump in yield. On the strength of experimental findings and experience, many rice scientists particularly at IRRI still believe that rice yield could be raised by 20-25 percent through selective improvement of major yield components.

A rice ideotype with a potential yield of 13-15 tonnes/hectare under irrigated situation has been designed with the following morphological and physiological features :

Morphological Features

(1) Semidwarf habit (90 cm tall) with stiff stem and with dark green, short, thick, erect leaves.

(2) Medium maturity duration (100 to 130 days).
(3) 3-4 large panicles with high grain number (200 to 250 grains/panicle) and with high grain weight (more than 25 g for 1000 grains); no unproductive tiller.
(4) High tillering ability under transplanted condition but low tillering ability (4 to 5 tillers) under direct seeding.

Physiological Features

(1) Faster foliar growth during crop establishment with reduced tillering.
(2) Slower foliar growth and enhanced translocation of assimilates from foliage to stem during late vegetative and reproductive stages.
(3) Expanding capacity of stem to store assimilates.
(4) Increased harvest index (usually 0.6 or more).
(5) Improved starch accumulation capacity in grain along with longer ripening period.
(6) Descending concentration of N from top to bottom in the crop canopy.

Physiological Needs for Higher Yield

Depending on the knowledge of crop photosynthesis, mineral nutrition and yield components, Yoshida (1981) suggested the following physiological needs for achieving high yields of rice:

(1) The plant must have a short and stiff stem.
(2) Erect leaves must be present at upper canopy levels whereas droopy leaves at low canopy levels.
(3) A LAI (leaf area index) of 5 to 6 is essential for maximum photosynthesis during reproductive stage.
(4) During grain filling stage, increased number of active green leaves must be maintained.
(5) Planting time should be so chosen that the reproductive or ripening stage, whichever is more critical to yield, is exposed to high solar radiation.

(6) All essential nutrients must be supplied through fertilisers to meet the crop's requirements.
(7) During ripening process, about 70% of the nitogen absorbed by the straw is translocated to grain.

It is essential to maintain high level of leaf nitrogen to produce high yields either by nitrogen absorption after flowering or by supplying already absorbed nitrogen before flowering from vegetative parts.

Ideotype for Upland Ecosystem

Scientists at IRRI formulated the rice ideotype for upland areas (IRRI highlights, 1987). Such ideotype must have the following features :

(1) Intermediate plant height with moderate tillering to improve nitrogen responsiveness.
(2) Droopy lower leaves to shade out weeds and conserve soil moisture.
(3) Erect upper leaves for efficient light penetration.
(4) Deep thick root system for drought resistance.
(5) Blast resistance and acid soil tolerance.

Ideotype for High Density Grain

High density grains refer to the heavier grains in the panicle. Rice scientists at IRRI have formulated the ideal ideotype for producing high density grains (IRRI highlights, 1987). The essential features of the ideotype are mentioned below :

(1) Low tillering with only primary tillers developing.
(2) Large panicles to compensate for low tillering.
(3) Thick culm with more vascular bundles to lessen lodging, to support a larger panicle and to promote carbohydrate accumulation.

(4) Panicles with only primary branches to increase the percentage of filled grains.
(5) Larger pedicelled vascular bundle to promote transport of assimilates.
(6) Medium-size grains (about the size of IR8 grains) with less white belly.
(7) Erect, thick leaves to give better light distribution and higher photosynthetic rate.
(8) Higher photosynthesis under low photosynthetically active radiation, so that carbohydrate supply will not be limited during rainy season.
(9) Low maintenance respiration to increase net assimilation rate; higher shoot : root ratio might decrease root maintenance respiration.
(10) Medium growth duration; carbohydrate accumulation before heading can be used to produce larger panicles and heavier grains.
(11) Intermediate plant height ; a harvest index of 0.55 will make the plant lodging resistant, decrease maintenance respiration and provide optimum partitioning of carbohydrate to the grain.

Ideotype for Flood Prone Ecosystem

Under flood prone ecosystem the rice plant must have the following characteristics :

(1) Intermediate plant height with thick culm and well developed sclerenchymatous band in the culm.
(2) Long, narrow, medium-thick, dark green leaves with high sugar and protein content.
(3) Deep root system with ability to withstand short drought spell when flood water recedes.
(4) Higher and rapid photosynthesis under low light intensity of submergence condition.
(5) Higher cytochrome oxidase activity to carry out functional activity at low O_2 concentration.

(6) Higher tensile strength of the root, stem and leaves to withstand the mechanical force of flood water.
(7) Regeneration ability of the shoot after recession of flood.

Ideotype for Deepwater Ecosystem

The rice ideotype for deepwater ecosystem must possess some specific genetical and morphological traits *e.g.*

Genetical traits

(1) High photoperiod sensitivity with late maturity.
(2) Higher seedling vigour.
(3) Submergence tolerance.
(4) Stem elongation ability with rising water level.
(5) Increased grain dormancy.

Morphological traits

(1) Seed germination under submergence condition with rapid elongation of plumule for direct seeding.
(2) Long, thin, droopy leaves with pubescence.
(3) Few tillers with nodal tillering ability at the water level.
(4) Rapidly spreading tillers with kneeing ability.
(5) Culm with elongated internode, and air space for floating (aerenchymatous tissue).
(6) Shallow and thin basal roots as well as nodal rooting ability.
(7) Only 1 to 4 photosynthetically active leaves at maturity.

REFERENCE

Yoshida, S. 1981. *Fundamentals of Rice Crop Science.* IRRI, Philippines.

12

Rice Yield Improvement

Introduction

Rice yield is influenced or modified by genetic characteristics of the variety, environmental factors and cultivation techniques over the period from seeding to harvesting. Yield is a complex polygenic trait with low heritability. Direct selection for yield is not economically rewarding.

Yield consists of several yield components *viz.* seedlings per hill, number of panicles per seedling, number of grains per panicle, ripening percentage and grain weight. Many rice workers have endeavoured to raise crop yields through the application of the concept of yield components that was proposed for plant breeding by Engledow and Badham (1923). The studies have resulted in small morphological changes in plant type and hence little change in yield because yield components show more or less inverse correlation between them (Watson, 1952). However, the yield components concept is still viable among the quantitative geneticists for analysing the results of rice cultivation after harvesting. Rice workers still estimate correlations among yield components and analyse direct effects and indirect effects of yield components on yield by path analysis.

Such analysis is retrospective (Sasahara, 1994) and may result in invalid data for next year's cultivation due to the fact that climatic conditions change from year to year (Tsunoda, 1958). Hence it is imperative that such analysis must be repeated over years to test the uniformity in results for better understanding of the magnitude of influence of yield components on yield.

Yield Increase in Japan

Increase in rice yield in Japan has been achieved through research efforts to prevent the overgrowth of vegetative parts (Matsuo, 1940a and b) in rice plant during 1930-1960. Overgrowth of vegetative parts reduces rice yield. It occurs due to two factors :

(a) cultivation of panicle weight type variety having tall stems susceptible to lodging.
(b) high fertilizer application before and/or at the time of transplanting(Sasahara, 1987).

But periodical fertilizer application between transplanting and the late heading stage reduces the overgrowth of vegetative parts.

Development of Ideal Plant-type

The ideal plant-type must possess erect, thick and dark green narrow leaves and short culms. A single plant is expected to have 6-10 culms — all of them productive — which bear large panicles, each with 200-250 grains. Thicker and sturdier culms support the larger panicles and with vigorous roots, prevent the plant from lodging. Thicker, dark green and erect leaves catch more sunlight and use it more efficiently. Short culms lower the consumption of assimilates (Sasahara, 1994).

The ideal plant-type can be developed in two ways:

(1) **Agronomic practice** : Agronomic practice, especially manipulation of fertilizer application, can produce an ideal plant-type. A reduction in nitrogen content in rice plants ca. 30 days before heading is intrinsically associated with development of ideal plant type at grain filling stage (Matsushima, 1957).

The nitrogen reduction in the plant at this growth stage results in prevention of the overgrowth of the flag leaf, the penultimate 2 to 3 leaves and the lower internodes and hence favours establishment of the ideal plant type. But the level of reduction in nitrogen content in the plant should be limited to allow formation of enough grains per unit land area to ensure a high rice yield.

(2) **Breeding approach** : The ideal plant-type has been developed by rice breeders through the development of short-statured plants having semi-dwarfing genes (DGWG gene). The semi-dwarf genes decrease the overgrowth of lower internodes, and of the flag leaf and penultimate leaves (Futsuhara, 1968). Exploitation of semi-dwarf genes has coincided with the ideal plant-type concept proposed by plant breeders (Tsunoda, 1959).

Yield and Solar Energy

Yield is represented by $Y = (SE \times EUE \times DGF)\, HC^{-1}$, where SE is the solar energy (cal cm^{-2} day^{-1}), EUE is the energy utilization efficiency, DGF is the days for grain filling, and HC is the heat of combustion (cal g^{-1} dry weight). SE cannot be changed at a given cultivation area and HC is a physical constant. DGF is also a constant. Only EUE can be increased to raise the yield.

An increase of EUE has been achieved through a high fertilizer application using lodging-tolerant semi-dwarf rice cultivars resulting in increased levels of chlorophyll, ribulose 1,5-biphosphate carboxylase (RuBiSCO) and other photosynthetic enzymes (Sasahara, 1994).

EUE ranges from ca. 1 per cent or less in C_3 plants to ca. 5 per cent in C_4 plants. C_4 plants are photosynthetically more active than C_3 plants. The efficiency of C_3 plants is limited by photorespiration which is absent in C_4 plants. Incorporation of the high photosynthetic characteristics of a C_4 plant (*Echinochlola oryzicola* L.) into C_3 rice plant (*Oryza sativa* L.) was attempted through protoplast fusion (Terada *et al.*, 1987), but the regenerated plant died. This may be due to the fact that the leaf tissue structures and enzyme construction of C_3 plants differ from those of C_4 plants, indicating the difficulty of reconstruction of the internal structures of leaves.

C_4 pathway is also called Hatch-Slack pathway after the name of two Australian biochemists, M.D. Hatch and C.R. Slack, who discovered it in 1960s. C_3 pathway is also called Calvin pathway after Melvin Calvin of U.S.A. who discovered it in 1940s. Both C_4 and C_3 plants ultimately use the C_3 pathway. The important difference is that in the C_4 plants, the C_3 pathway is preceded by additional steps in which there is a preliminary fixation of CO_2 into a compound with four carbon atoms before the CO_2 is incorporated into the three carbon compound phosphoglycerate (Lehninger, 1984). C_4 plants need five ATPs to fix one molecule of CO_2 whereas the C_3 plants need only three. C_4 plants are of tropical origin and grow faster and make more biomass per unit leaf area than the C_3 plants of temperate zones.

In C_4 plants, phosphoenol pyruvate carboxylase enzyme which fixes CO_2 to oxaloacetate is located in mesophyll cells and is the most active agent for absorbing CO_2. But the activity of RuBiSCO present in the vascular bundle sheath cells of C_4 plant is similar to or less than that in the mesophyll cells of C_3 plant. Therefore, attempts to increase EUE in the C_3 rice plant will require further experiments.

Structural Improvement of Organ

Rice yield has been raised through the structural improvement of organs including culms and leaves and also through the increase in N-containing ingredients such as chlorophyll and RuBiSCO through high fertilizer application. Yield has been increased through the replacement of panicle weight type by panicle number type rice plant.

The future goal for raising rice yield appears to be qualitative improvement at the tissue and molecular levels, including increase in the activity of RuBiSCO. The latter may result in a reduction of photorespiration, a wasteful process, and hence a decrease of light saturation in C_3 plants. These advancements would be achieved through the development of biotechnology and genetic engineering in rice plant.

CO_2 Diffusion

CO_2 in the turbulent air diffuses through the leaf stomata to finally reach the chloroplast. CO_2 diffuses through several resistances *viz.* resistance of boundary air layer covering the leaf, cuticle resistance, stoma resistance and mesophyll resistance. Of these, cuticle resistance has a very large value.

The diffusion rate of CO_2 depends upon the molecular concentration gradient between the turbulent air and the chloroplast. The CO_2 concentration in the turbulent air is estimated to be ca. 0.330 ml/L and is hypothesised to be zero in the chloroplast. By increasing the activities of RuBiSCO and other photosynthetic enzymes, more severe starvation of CO_2 at the chloroplast may be created. It will help in greater diffusion of CO_2 for increased photosynthesis, thereby raising the yield.

Improvement of Leaf Morphology

Development of V-type leaf through breeding has been found to increase rice yield in Japan (Sasahara, 1994). The angling of leaf blade decreases the light interception and hence the light extinction coefficient (K) in a plant community. This brings about a high light intensity at lower parts in the plant, resulting in a high physiological activity of leaves at the lower layers and hence an increase in the transport of assimilates into roots. The still air layer may become thinner in theV-type leaf than in the flat leaf commonly found, resulting in a high rate of CO_2 diffusion.

Light extinction coefficient (K) mentioned above is an item in the Beer-Lambert's law which is given by Log $(I_i\, I_0^{-1})$ = -KF where Log$(I_i\, I_0^{-1})$ is the relative light intensity at a given depth in a solution, K is the light extinction coefficient and F is the concentration of colouring matter dissolved in the solution. According to this law, K is completely dependent on F. The concept of light extinction

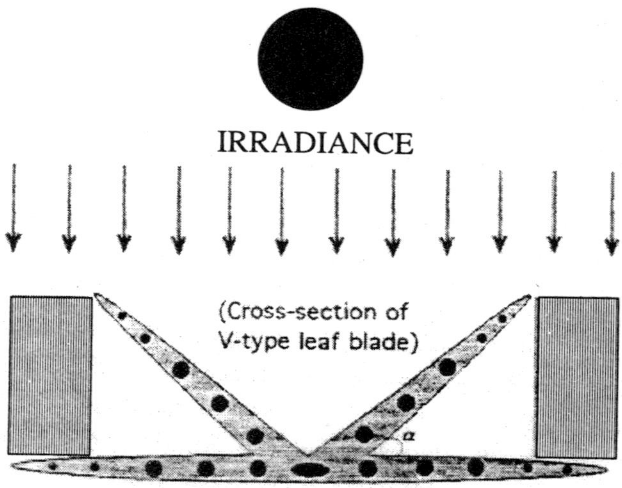

(Cross-Section of flat leaf blade)

● : Mid-rib or vein

▥ : No light interception area induced by leaf blade tilting.

Courtesy: Sasahara (1994)

Fig. 12.1 :Schematic representation of light interception in flat and V- type leaves.
Note: Light interception in the V-type leaf decreased in proportion to cosine *a*

in Beer-Lambert's law could be adopted for the plant canopy and the light extinction coefficient (K) could be changed between plant communities even at the same level of F. This was attributed to differences in size, angling, shape and thickness of leaves. This concept was adopted for the improvement of the plant type in several crops by Tsunoda (1959).

A preliminary yield test with the line DARM-7-2 possessing V-type leaf yielded ca. 10 tonnes/ha with a high fertilizer application (Sasahara, 1994).

Low Yield of Indica Rice in Tropics

Farmers in the tropics grow traditional rice varieties which are poor responsive to nitrogen. When nitrogen is applied at rates exceeding 60Kg/ha these varieties grow tall, tiller profusely, lodge early and produce low yields (Chandler, 1963). Lack of nitrogen responsiveness stands as a barrier for increasing the yield of indica rices in the tropics. Low performance of indicas in the tropics is attributed to:

(1) warm climate in the tropics.
(2) genetic association of undesirable plant characters.

Traditional tropical indica varieties are late maturing, photoperiod sensitive and high tillering with tall, weak and thick culms. They possess long drooping, pale-green, thin leaves and relatively large lax panicles. These are grown at wide spacing under poor weed and water management.

On the other hand, subtropical japonica rices which are highly responsive to nitrogen are high yielding. They have thin, but short and sturdy culms with short, narrow, erect, dark-green leaves and short, dense panicles. Such observations suggest that japonica rices be grown in the tropics. But the direct introduction of japonica rices in the tropics was not favoured for several reasons :

(1) Most japonica rices mature very early in the tropics due to their daylength and temperature sensitivity, resulting in stunted growth and very low yield.
(2) Japonica rices lack seed dormancy desirable for tropical varieties that mature towards the end of monsoon.
(3) People in the tropical Asia prefer the cooking quality of indica rices and not of japonicas.

Considering the above factors attempts were made to make hybrids between indicas and japonicas in order to transfer good characters of japonica rices to indica rices. In the 1950s, the FAO organised an indica-japonica hybridization programme with its headquarters at CRRI, Cuttack. As a result, 'ADT 27' from India and 'Mahsuri' and 'Malinja' from Malaysia came out from this programme.

The warm climate in the tropics was considered responsible for low yield of rice. Murata (1964) correlated the rice yields with solar radiation and air temperature by the following formula:

$$Y = S\,[1.20 - 0.021(t - 21.5)^2]$$

Where Y is grain yield in kg/0.1 ha, and S and t are solar radiation (cal/cm^2 per day) and daily mean temperature (^{o}C) in August and September. The formula suggests that temperatures higher or lower than 21.5^{o}C will decrease grain yield. In general, high temperatures increase respiratory losses and reduces net dry matter production, finally lowering the yield.

Subsequent work at the IRRI demonstrated that high temperature *per se* in tropical monsoon Asia is not a barrier to increased rice production upto 5-6 t/ha in the wet season and 9-10t/ha in the dry season. It was the variety that hampered attempts to increase rice yields in experimental fields.

Shuttle Breeding

Rice is grown under diverse environments throughout the world. It is subjected to different stresses (both biotic and abiotic) *viz.* drought, flood, high and low temperatures, soil acidity and alkalinity, disease infection and insect-pest attack.

In order to develop improved rice varieties for many a stress situation, early generation segregating materials are exposed to the actual stress environment for effective selection. Most of the stress situations do not occur at International Rice Research Institute in Los Banos, Philippines. For more rapid progress, IRRI scientists have developed shuttle breeding in collaboration with the rice researchers in localities/countries where the adverse stresses actually occur (IRRI highlights, 1986).

In this breeding approach, shuttling between two locations (IRRI and target locality) for growing alternate generations of rice breeding lines facilitates selection for stress tolerance in the target environment as well as for grain quality and disease and insect

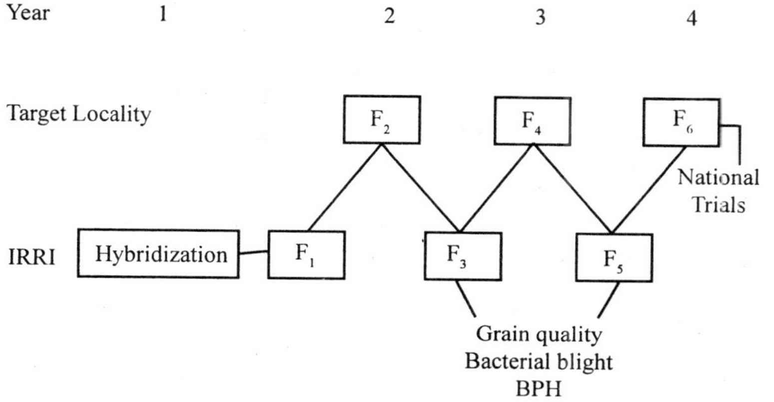

Fig. 12.2: Shuttle Breeding

resistance.

Parents that contribute the desired resistance or tolerance to a cross (usually an improved variety or breeding line from IRRI and a locally adapted variety of the collaborating country) are selected jointly by IRRI and national scientists. Crosses are made and the F_1 progenies are grown at IRRI. The F_2 populations are grown in the target environment, where they are exposed to the appropriate stress condition.

Seeds of locally adapted plants selected from the F_2 population are returned to IRRI for evaluating F_3 progeny resistance to diseases and insects and for grain quality. Selected F_4 progeny are again grown in the target environment and the F_5 at IRRI. IRRI took up shuttle breeding projects for developing breeding materials with cold tolerance and with adaptability to rainfed lowland, tidal wetland and deep water conditions.

REFERENCES

Chandler, R.F. Jr. 1963. An analysis of factors affecting rice yield. International Rice Comm. *Newsl.*, 12(4): 1-17.

Engledow, F.L. and Badham, S.M. 1923. Investigation on yield in the cereals. *Indian J. Agric. Sci.*, 13; 390-349.

Futsuhara, Y. 1968. Breeding a new rice variety "Reimei" by gamma ray radiation. Gamma Field Symposia, 7:87-109.

Lehninger, A.L. 1984. *Principles of Biochemistry.* CBS Publishers and Distributors, Delhi.

Matsuo, D. 1940a. Relationship of differences of developmental phase during plant growth to yield, especially to grain quality. 1. *Agric. Hort.,* 15:845-852.

Matsuo, D. 1940b. Relationship of differences of developmental phase during plant growth to yield, especially to grain quality. 2. *Agric. Hort.,* 15:1055-1059.

Matsushima, S. 1957. Analysis of developmental factors determining yield and yield prediction in lowland rice. *Bull. Natl. Inst. Agr. Sci.* Ser. A5:1-271.

Murata, Y. 1964. On the influence of solar radiation and air temperature upon the local differences in the productivity of paddy rice in Japan. *Proceedings Crop Sci. Soc.* Japan, 35:59-63.

Sasahara, T. 1994. A new strategy for selecting high yielding rice. Gamma Field Symposia, Institute of Radiation Breeding, NIAR, MAFF, Japan, 33:19-34.

Sasahara, T. 1987. Relationship of photosynthesis and plant production to yield. *Agric. Hort.,* 62:463-471.

Terada, R. Kyozuka, J., Nishibayashi, S. and Shimamoto, K. 1987. Plantlet regeneration from somatic hybrid of rice (*Oryza sativa* L.) and barnyard grass (*Echinochlola oryzicola* Vasing). *Mol. Gen. Genet.,* 210:39-43.

Tsunoda, S. 1958. Relationship of the high yielding ability of crops to their growth characteristics. *Agric. Hort.,* 33:1469-1474.

Tsunoda, S. 1959. A developmental analysis of yielding ability in varieties of field crops. II. The assimilation system of plants as affected by the form, direction and arrangement of single leaves. *Japanese J. Breeding,* 9:237-244.

Watson, D. J. 1952. The physiological basis of variation in yield. *Adv. Agron.,* 4:101-145.

13

Enhancing Genetic Yield Potential in Rice

Introduction

The area under rice cultivation worldwide has remained almost same since 1980. In the coming decades also, major increase in the area under rice is unlikely mainly due to the pressures of industrialization and urbanization. In order to feed the ever increasing human population worldwide, the rice production has to be increased by increasing the productivity, rather than by increasing the area under cultivation.

The world population in 1998 was 5.8 billion, it is likely to reach 7 billion in 2010 and 8 billion in 2020. As compared to the present world production of rice, nearly 350 million tons additional rice will have to be produced by 2020. Rice production will have to maintain the 3% average annual growth rate achieved over the last 20 years.

During the last three decades, rice production increased at a slightly higher rate than the population. This has saved the human population from starvation. But it is a fact that the rate of increase in rice production is slowing down and if the trend is not reversed, severe food shortage will occur in the beginning of next millennium. The population of rice consuming countries is increasing at a faster rate than that of the rest of the world. The number of rice eaters will probably increase by 70% during the next 20 years.

World's rice production during last 3 decades has increased due to the cultivation of high-yielding semi-dwarf rice varieties with improved technology. The increased demand for rice in the years ahead has to be met from less land with less water, less labour and less pesticides. So the need of the time is to develop rice varieties with higher yield potential to meet the goal of increased rice production.

Increasing Yield Potential

Yield potential of present day high yielding semi-dwarf rice varieties under the best conditions in the tropics is 10 tons per hectare during dry season and 6.5 tons per hectare during wet season (Khush, 1996). Plant physiologists have suggested that the physical environment in the tropics is not a limiting factor to increase rice yields.

Yoshida (1981) estimated the maximum yield potential of rice as 15.9 tons and 9.5 tons per hectare during the dry season and the wet season, respectively. Maximum rice yield potential can be realised through two approaches of conventional breeding. These approaches are :

A. Development of New Plant Type
B. Exploitation of Hybrid Vigour

A. Development of New Plant Type

Development of the plant type of modern high yielding semi-dwarf rice varieties in the mid-1960s almost doubled the rice yield. This plant type is short statured and possesses sturdy stems, dark green erect leaves and high tillering ability. Number of tillers in this plant type varies from 20 to 25 with nearly 14 to 16 tillers bearing panicles but the remaining tillers are unproductive. More than 60% of world's rice area is now under varieties of this semi-dwarf plant type.

During the last 30 years, no significant increase in yield potential was achieved. This is because rice improvement efforts at IRRI were directed towards incorporation of disease and insect resistance, early maturity and improved grain quality in semi-dwarf varieties. For another quantum jump in rice yield, the yield potential must be enhanced either by increasing the harvest index (HI) or biomass (total dry matter production) or both in a new plant type.

Increasing Harvest Index (HI)

Harvest index is defined as the ratio of economic yield (grain) to biological yield (straw) excluding the roots. In other words, harvest index is the grain to straw ratio. The HI of modern high-yielding cultivars varies from 0.45 to 0.53 depending upon the season. The HI is higher during dry season than that of wet season (Vergera and Visperas, 1977). Harvest index should be increased to around 0.60 in the rice varieties with new plant type for increased yield potential. The maximum possible HI in wheat was estimated to be 0.63 (Austin *et al.*, 1980). The HI of 0.63 may also be considered as the maximum harvest index for rice assuming the same biomass production of the modern semi-dwarf plant type.

For increasing the harvest index, the following plant characteristics need improvement (Khush, 1994) :

(1) Increased sink size
  (a) large spikelet number per panicle.
  (b) greater partition of assimilate in spikelet formation

(2) Increased spikelet filling
  (a) reducing canopy senescence
  (b) higher proportion of high-density (heavy weight) grains
  (c) maintaining healthy root system
  (d) increased lodging resistance.

Increasing Biomass

Yield is a function of biomass or total dry matter production. Increasing biomass production will increase rice yield. Biomass production can be increased by growing rice crop in a high solar environment similar to that of dry season in the tropics provided there is sufficient nitrogen supply to the rice crop (Akita, 1989). In general, increased biomass production produces tall culms which result in lodging and mutual shading. As a result of mutual shading, humidity in the lower canopy increases and it helps in increasing disease incidence consequently reducing grain yield (Vergara, 1988). If lodging and disease problems can be solved, increased biomass production could increase yield potential in tropical environments.

Biomass production can be enhanced by improving the following plant characteristics:

(1) Establishment of desirable canopy structure
  (a) rapid leaf area development
  (b) rapid nutrient uptake

(2) Reducing carbon consumption.

Phenotype of New Plant Type

For achieving high harvest index and high biomass production, a new plant type (ideotype) has been conceived with the following characteristics (Khush,1994):

(1) low tillering capacity (3-4 tillers for direct seeding and 6-8 tillers under transplanting)
(2) no unproductive tillers
(3) 200-250 grains (spikelets) per panicle
(4) 90-100 cm tall
(5) very sturdy stems
(6) dark green, thick and erect leaves
(7) vigorous root system
(8) 100-130 days growth duration (i.e seed to seed duration)
(9) acceptable grain quality
(10) multiple disease and insect resistance.

Distinct Plant Features for Increased Yield Potential

1. **Plant height and stem thickness**: A decrease in plant

height reduces total biomass production and resists lodging. But short culms increase harvest index and require less maintenance respiration. Reduced respiration of short culms leads to an increased balance between photosynthesis and respiration (Tanaka *et al.*, 1966) which results in more storage of photosynthate. A plant height of 90-100 cm is considered ideal for maximum yield. Plant height less than 90 cm considerably reduces total biomass, thereby reducing the grain yield. Thicker stems accumulate more assimilates and resist lodging. Thicker stems also have more vascular bundles which enhance photosynthate translocation.

2. **Low tillering and large panicles** :Modern semi-dwarf rice varieties produce nearly 20-25 tillers under favourable environment but of these, only 14-15 tillers bear panicles. Other tillers are unproductive and compete with the productive tillers for photosynthates, solar energy, mineral nutrients (particularly nitrogen) and space. High tillering produces a dense canopy which restricts the penetration of sunlight and creates a shady humid microenvironment favourable for disease, particularly for the endogenous pathogens like sheath blight *(Rhizoctonia solani)* and stem rot *(Sclerotium oryzae)*.

   Elimination of unproductive tillers through breeding will direct more nutrients for the grain production resulting in increased yield. Low tillering genotypes are characterised by synchronous flowering and maturity and more uniform panicle size which help in harvesting. The proportion of high density grains (heavy weight grains) is high in a low tillering genotype.

   High tillering produces some unproductive tillers which lead to high leaf area index (LAI) and vegetative growth but produces more unfilled grains. The new plant type must have a few tillers, each with large panicle. Traditional rice varieties possess fewer tillers than the modern semidwarf varieties. Large panicle will compensate for small number of panicles in a low tillering genotype.

3. **Grain characteristics**: The proportion of heavier, high density grains is the greatest at the top of the panicle and the smallest in the inferior spikelets of the lower panicles. A thousand grain weight of 25 g is ideal for rice. Larger grains (*i.e.* higher thousand grain weight) are chalky and thus have lower market value. Large variation in the weight of single rice grains was reported by Venkateswarlu *et al.* (1986).

   The proportion of high density grains is higher in the primary tiller than in the secondary or tertiary tiller (Rao, 1987b) . Low tillering genotypes produce more primary tillers per unit land area and hence more high density grains per unit land area. High density grains contribute to increased yield potential.
4. **Root system :** Thicker and deeper roots are desirable in an ideal plant. These roots absorb water and minerals more efficiently from soil. In addition to this, the deeper roots provide better anchorage of rice plant to soil and reduce lodging. Healthy roots are necessary during grain filling period for better nutrient supply. Variation in rice varieties for root characteristics namely root length, root volume, root thickness and degree of branching has been observed. Such variation could be exploited for increasing rice yield potential.
5. **Growth duration** : Seed to seed growth duration of 120 days is considered ideal for maximum yield potential of rice in the tropics. Variation in growth duration among rice varieties is mainly due to the difference in the vegetative growth period (Vergera *et al.*, 1969). Studies have also indicated that panicle growth period (from panicle initiation to heading) also differs in early-maturing and late-maturing varieties. In general, an early maturing variety has relatively short panicle growth period than a late-maturing variety. Short panicle is often accompanied by decreased grain yield. Thus, increasing yield potential may be explored by extending the panicle growth period and also

possibly by increasing the grain filling period independent of the total growth duration (Khush, 1996).

6. **Leaf and canopy characteristics** : In a high yielding variety, the leaf must be small, thick, erect and dark green in colour. V-shaped leaf is preferable to flat leaf for increased photosynthesis due to improved CO_2 penetration. Light is used more efficiently at a high leaf area index (LAI) in an erect-leaved canopy (Yoshida, 1976). Carbon assimilation of a single leaf exposed to light on only one side is lower than when the leaf is exposed to light on both sides and this difference is the greatest when leaves have a high nitrogen content and greater thickness.

   Long, thin leaves of the traditional rice varieties are droopy or horizontally oriented. These leaves reduce sunlight penetration and air movement. As a result, the canopy becomes shady and humid. High humidity, in turn, favours the growth of endogenous pathogens like sheath blight and stem rot, consequently decreasing rice yield.

   Thicker leaves are not associated with yield potential but positively correlated with leaf photosynthetic rate. Thicker leaves are thought as desirable trait and can be visually selected with ease as a criterion for enhancing yield potential.

7. **Disease and insect resistance**: The prospective new plant type with increased yield potential adapted to tropical conditions must have resistance to major diseases like blast, bacterial blight, sheath blight and several viral diseases and also to major insects like stem borers, leaf hoppers and plant hoppers (Khush, 1996).

B. Exploitation of Hybrid Vigour

Exploitation of hybrid vigour through development of hybrid variety is another approach for enhancing the genetic yield potential in rice. The magnitude of hybrid vigour (or heterosis) depends on the genetic diversity of the two parents of a hybrid. The greater the genetic difference between two parents the higher the hybrid vigour.

During the past 30 years, the genetic diversity among the improved indica lines has narrowed down due to large international exchange of germplasm (Khush and Aquino, 1994). Indica and japonica germplasm have remained distinct as there has been very little gene flow between these two varietal groups. Experimental results have shown that the magnitude of heterosis in indica/japonica hybrid is 10 to 15% higher than indica/indica hybrid and 20 to 25% higher than japonica/japonica hybrid.

Attempts are being made to develop new plant type in the tropical japonica (or javanica) germplasm. This new plant type will be used in hybridization with indica rice to produce hybrids with high heterosis. The indica/japonica hybrids are expected to have 20–25% yield advantage over the best inbred lines in farmers' field whereas the new plant type in the tropical japonica background is expected to have a yield advantage of 20% over the existing high yielding indica rice varieties.

Progress in Breeding

Research work for breeding the new plant type was initiated in 1989 at IRRI. About 2000 entries from IRRI germplasm bank were grown in dry and wet seasons to identify donors for various traits. Hybridization work was started in the 1990 dry season and F_1 progenies were grown in the 1990 wet season. F_2 progenies were raised in the 1991 dry season and a pedigree nursery in the 1991 wet season. By 1996, nearly 1350 crosses were made and ca. 60,500 pedigree nursery rows were grown.

Breeding lines with the desired characteristics of the new plant type were selected and subjected to replicated yield trials. Several new plant type lines were evaluated along with IR 72, the improved indica variety and are in the pipeline for release. Further research work will be directed towards improvement of disease and insect resistance and also grain quality.

In order to develop the new plant type, javanica germplasms were examined by the breeders at IRRI. Javanica varieties are also called bulu rice which is native to Indonesia. Javanicas are low

Table 13.1: Donors for various traits being used for developing the new plant type

Trait	Donors
Short stature	MD2, Sheng-Nung 89-366
Low tillering	Merim, Gaok, Gendjah Gempol, Gendjah Wangkal
Large panicles	Daringan, Djawa Serang, Ketan Gubat
Thick stems	Sengkeu, Sipapak, Sirah Bareh
Grain quality	Jhum Paddy, WRC4, Azucena, Turpan4
Donor for resistance	
Bacterial blight	Ketan Lumbu, Laos Gedjah, Tulak Bala
Blast	Moroberekan, Pring, Ketan Aram, Mauni
Tungro	Gundil Kuning, Djawa Serut, Jimbrug, Lembang
Green leafhopper	Pulut Cenrana, Pulut Senteus, Tua Dikin

After Khush (1996).

tillering with large panicles and sturdy. These characteristics were looked for by the breeders for developing new plant type.

Javanicas are genetically very close to japonicas grown in the temperate regions. On the basis of similarity observed in isozyme analysis at 15 loci, it was found that javanicas and temperate japonicas belong to the same japonica varietal group. Since javanicas are grown in the tropics, they are known as tropical japonicas. Crosses between temperate japonicas and tropical japonicas are fully fertile and there are no barriers to recombination. But crosses between japonicas and indicas have varying levels of sterility and produce poor recombinants because genetic barriers to recombination exist in such crosses.

For developing new plant type, breeders at IRRI decided to restrict hybridization work within the tropical germplasm with selective introduction of genes from indicas and temperate japonicas. This decision was based on 3 reasons namely (Khush, 1996):

(1) Crosses within tropical japonicas would not face the

problems of sterility and restriction to recombination.

(2) F_1 hybrids between indica and japonica rices are highly heterotic. But temperate japonicas are not adapted to tropical conditions and it is very difficult to produce indica/japonica hybrid. For this purpose, improved tropical japonicas with genes for wide compatibility, short stature, disease and insect resistance and long slender grains would be most suitable (Khush and Aquino, 1994).

(3) Since the development of modern semi-dwarf rice varieties in the mid-1960s by introducing dwarfing genes in indica germplasms, no significant breakthrough in rice yield has been made by IRRI and different national programmes despite continued research efforts. So the breeders at IRRI decided to work with entirely different germplasm of tropical japonicas for enhancing the yield potential of rice.

REFERENCES

Akita, S. 1989. Improving yield potential in tropical rice. Progress in irrigated rice research. Proceedings International Rice Research Conference, Sept., 21-25 IRRI, Philippines, 41-73.

Austin, R.B., Bingham, J., Blackwell, R. D., Evans, L.T., Ford, M.A., Morgan, C. L. and Taylor, M. 1980. Genetic improvements in winter wheat yields since 1990 and associated physiological changes. *J. Agric. Sci.*, Cambridge, 94:675-689.

Khush, G.S. and Aquino, R. 1994. Breeding tropical japonicas for hybrid rice production. Hybrid rice technology, new development and future prospects. IRRI, Philippines, 33-36.

Khush, G.S. 1996. Prospects of and approaches to increasing the genetic yield potential of rice. Rice Research in Asia : Progress and Priorities, CAB International & IRRI, Philippines, 59-71.

Rao, S.P. 1987b. High density grain among primary and secondary tillers of short and long duration rices. *International Rice Research Inst. Newsl.*, 12 (4):12.

Tanaka, A., Kawano, K. and Yamaguchi, J. 1966. Photosynthesis, respiration and plant type of the tropical rice plant. *Technical bulletin number* 7, IRRI, Philippines.

Venkateswarlu, B., Parao, F.T., Visperas, R.M. and Vergara, B.S. 1986. Screening quality grains of rice with a seed blower. *Society for the Advancement of Breeding Researches in Asia and Oceania J.*, 18(1): 19-24.

Vergara, B.S., Chang, T.T. and Lillies, R. 1969. The flowering response of the rice plant to photoperiod. *Technical bulletin number* 8, IRRI, Philippines.

Vergara, B. S. and Visperas, R. M. 1977. Harvest index : criterion for selecting rice plants with high yielding ability. IRRI Saturday seminar, September 10, IRRI, Philippines.

Vergara, B.S. 1988. Raising the yield potential of rice. *Philippines Technical J.*, 13:3-9.

Yoshida, S. 1976. Physiological consequences of altering plant type and maturity. Proceedings of International Rice Research Conference, IRRI, Philippines, 268.

Yoshida, S. 1981. *Fundamentals of rice crop science.* IRRI, Philippines.

14

Rice Seed Production and Testing

Introduction

Seed is the basic input of crop production. Use of high quality seed is the most effective and the cheapest means of getting highest return from cultivation. Good quality seeds result in uniform germination and growth. High quality seeds possess some common characteristics *e.g.*

(1) high germination percentage,
(2) high degree of analytical purity,
(3) uniform and correct size,
(4) freedom from seed-borne diseases.

Classes of Seeds

To ensure the supply of high quality rice seeds to the needy farmers, different classes of seeds are produced. Knowledge of the different seed classes is important not only for the breeders but also for the farmers.

(1) **Breeder Seed:** It is the first step in seed production. Breeder seeds are controlled by the originating or sponsoring plant breeder or institution. These seeds are not available to the public. White tags are affixed to breeder seed packets.

(2) **Foundation Seed**: Foundation seed is the second step in seed production programme. Foundation seeds are the progeny of the breeder seed. The plant breeding institution or NSC (National Seed Corporation) which developed the particular rice variety has the responsibility to maintain and produce foundation seeds. They act as the source for the production of registered seed or certified seed. Red tags are affixed to foundation seed packets. Foundation seed is produced on Government farms, Agricultural Universities or on farmer's field under strict supervision of research scientists and experts from NSC.

(3) **Registered Seed**: These seeds are the progeny of breeder or foundation seeds. These are properly handled to maintain the genetic identity and purity. The selected seed growers produce and maintain registered seeds. Green tags are used to designate this class of seeds. In India, registered seeds are not produced.

(4) **Certified Seed**: Certified seeds are the progeny of the breeder, foundation and registered seeds which are properly handled to maintain genetic identity and purity. This class of seed is used by farmers for cultivation. Blue tags are affixed to certified seed packets.

Purity and Quality Standards

The farmers must consider the physical purity, moisture content of the seed and its germinability before buying seeds in order to reap a good harverst (Table 14.1).

Seed Purity

Seeds must be physically pure. The seed of one variety must not be mixed with seeds of other varieties. Seed purity analysis is done in the laboratory and component parts are separated into—pure seed, inert matter, other crop seed and weed seed.

Pure seed refers to the variety predominant in the test sample.

Table 14.1: Laboratory seed standards for rice

Factor	Standard for each class		
	Breeder	Foundation	Certified
Pure seed (%) (minimum)	98	98	97
Weed & other crop seed (%) (max.)	0	0	0.10
Inert Matter (%) (max.)	2	2	3
Red rice (grain / 500 grams) (max.)	0	0	2
Other varieties (grains/ 500 grams) (max.)	0	2	10
Germination (%) (min.)	80	80	80
Moisture Content (%) (max.)	14	14	14

Inert matter includes seeds that are very small or undersized or broken, seed-like structures and other matters like soil, sand, stem, leaves or dirt. Other crop seed includes seeds and seed-like structure of any plant species other than that of pure seed. Weed seed refers to seed and seed-like structures of weed.

Moisture Content

Moisture content of seed is the critical determinant of seed longevity and grain quality during storage. Seeds with 13 per cent moisture can be stored for one year without compromising viability. If seed is to be kept for two years, the moisture content should be 12-13 per cent. A high moisture content shortens longevity and thereby reduces its quality.

Germinability

Seed germination test measures the viability or the ability of the seed to produce normal seedling under normal conditions. It provides information on the field planting value of the seed. Germination per cent recorded in the laboratory is usually reduced in field conditions due to unfavourable soil and weather conditions, pests and diseases and competition for nutrition.

Methods of Testing Seed Viability

There are different methods of testing seed viability. These are briefly described below :

(1) **Petri-dish Method**: Place a number of randomly selected seeds from the seed sample in a petri-dish and pour enough water to moisten the seeds. Cover the seeds with a filter paper or any paper that can retain moisture for days. Determine germination per cent after five days using the formula given below :

$$\text{\% Germination} = \frac{\text{No. of Seeds germinated}}{\text{No. of seeds used}} \times 100$$

If the germination is 80 per cent or above, the seed is ideal for sowing. If the germination per cent is between 60 and 80 per cent, the seed rate must be increased. The new adjusted seed rate is calculated as given below :

$$\text{Adjusted seed rate (Kg/ha)} = \frac{1}{\text{Germination percent}} \times \text{Original seed rate (kg/ha)}$$

Adjusted seed rate will be higher than original seed rate.

(2) **Seedbox Method**: Place fine sand in seed boxes. Moisten the sand to 50 per cent of its water holding capacity. Plant randomly selected seeds from the seed sample in uniform layers of moist sand. Cover to a depth of 10-20 millimeters

and water daily. After 5 days take the germination percentage.

(3) **Ragdoll Method :** Prepare a rag of convenient size and a stick. Take some seeds from the sample and distribute them evenly in rows in the rag. Then roll the rag in the stick for support. Moisten the cloth and place it in a warm, moist place and water daily. After five days unroll the cloth and then determine the germination percentage.

15

Genetic Markers in Rice

Introduction

A genetic marker is an easily detectable trait with simple inheritance. To be useful as a genetic marker, a trait must meet two criteria :

(1) the trait must differentiate between the parents, and
(2) the trait must be accurately reproduced in the progeny.

For example, blood type is often different between two parents and the parental blood types are accurately reproduced in the progeny, so blood type must be used as a genetic marker. In contrast, height is often different between two parents, but the parental heights are usually not accurately reproduced in the progeny, so height is generally not useful as a genetic marker (except in extreme cases, such as genetic dwarfism).

Types of Markers

In the construction of genetic maps, three types of markers namely morphological, biochemical and molecular markers are used :

Morphological Markers

Morphological markers are externally visible morphological traits. They have the following characteristics :

(1) Morphological markers are limited in number.
(2) They are largely influenced by the environment.
(3) They have pleiotropic effect (genes governing more than one trait).
(4) Morphological markers manifest dominance-recessive expression.
(5) They show differential expression across the genotypes.
(6) They are not distributed throughout the genome.
(7) Expression depends on the ontogeny of the individuals. In other words, expression is development stage specific (temporal variation). Morphological markers express at a specific developmental stage.

Biochemical Markers

Isozymes and proteins are the commonly studied biochemical markers. Isozymes are the multiple forms of an enzyme that catalyse the same biochemical reaction but differ in their kinetic properties. They have distinctly different amino acid composition or sequence and may occur in the same species, in the same tissue, or even in the same cell (Lehninger, 1984). They have different molecular forms and are encoded by multiple alleles of a gene (locus) or by a set of closely linked alleles of a gene cluster. The isozymes can be easily separated by electrophoresis in polyacrylamide or starch gels. Besides, the isozymes also differ in their substrate affinity, in their maximum activity, or in regulatory properties.

The biochemical markers have several features :

(1) Biochemical markers are limited in number.
(2) They are influenced by internal environment of an organism.
(3) They are developmental stage specific (temporal variation).
(4) They are also tissue specific (spatial variation).

Isozymes of several enzymes such as peroxidase, esterase, acid phosphatase, amylase etc. have been extensively studied in crop plants.

Molecular Markers

Molecular markers refer to the DNA markers. They possess distinct characteristics:

(1) Molecular markers are numerous.
(2) They show Mendelian inheritance and are stably inherited.
(3) They are not influenced by environment.
(4) They are codominant and do not behave in a dominance-recessive manner.
(5) They are distributed throughout the genome.
(6) Molecular markers do not show pleiotropic effect.
(7) They are detected at any growth stage or developmental stage of the organism.
(8) They find several applications in crop improvement through gene tagging, marker aided selection, gene pyramiding, QTL mapping etc.

The molecular markers include short DNA sequences, whole genes or even longer sequences of DNA. Since the study of molecular markers has immense potential in understanding the molecular biology of crop plants, it is pertinent to give a general account of molecular markers in the following discussion for the benefit of the readers. So this chapter will deal with the concepts, methods and applications of molecular markers with subsequent appropriate emphasis on rice.

Types of Molecular Markers

Several types of molecular markers are available namely :

(1) Restriction Fragment Length Polymorphism (RFLP)
(2) Randomly Amplified Polymorphic DNA (RAPD)
(3) Polymerase Chain Reaction (PCR) Based Markers viz.

(a) Sequence Tagged Site (STS)

(b) Microsatellite
(c) Amplified Fragment Length Polymorphism (AFLP)
(d) Sequence Characterised Amplified Region (SCAR)
(e) Variable Number Tandom Repeats (VNTR)

Out of these RFLP and RAPD are the most commonly studied molecular markers.

Objectives of Molecular Markers

The major objectives of molecular markers are :

(1) To tag genes for economic traits.
(2) To pyramid genes for developing disease and pest resistant cultivars.
(3) To increase the efficiency of plant breeding through indirect selection rather than screening of large progenies.

Restriction Fragment Length Polymorphisms (RFLPs)

RFLPs are genetic differences observable at the DNA level characterised by a number of variable length resrtricted fragments. Original documentation of RFLPs resulted from the findings of Grodzicker *et al.*(1974). The presence of RFLP was demonstrated in the globin genes of man.

The RFLP analysis involves DNA cleavage by restriction enzymes (see Box 1), electrophoresis of resulting fragments, southern transfer of separated fragments to a membrane support, radioactive labelling of suitable probes, hybridisation of probes to membrane supported fragments and detection as a banding pattern on X-ray film. RFLPs can be of various types : single copy sequences, multiple copy sequences and repeated sequences. Among these, single copy sequences are highly amenable for linkage map construction and to some extent, the multiple copy sequences (Maheswaran, 1997).

Box - 1

Restriction enzymes are a class of DNA endonucleases which are found in various microbes. Type II restriction enzymes cleave DNA at sites where a specific base sequence (restriction site) occurs and are commonly used in molecular biology. Restriction sites for various restriction enzymes vary from 4 to 8 base pairs length and are usually palindromic (reads the same from opposite directions). The restriction site for Pst I restriction enzyme is thus :

```
              ↓
5' – C T G C A  G – 3'

3' – G  A C G T C – 5'
       ↑
```

The arrows denote the site where the phosphodiester back bone of the DNA is cleaved.

Restriction enzymes are thought to have evolved as a defense mechanism against foreign DNA and degrade only the foreign DNA (such as that of bacteriophage) in a bacterial cell and not the DNA of the bacterial cell itself. Bacteria protect their own DNA from being degraded by attaching methyl groups to nucleotides in potentially sensitive restriction sites.

RFLPs and Genome Characterization

RFLPs have some advantages in genome characterisation :

(1) RFLPs show Mendelian inheritance.
(2) They give codominant expression.
(3) They lack pleiotropic effects.
(4) They are not influenced by environment.
(5) RFLPs are present at each developmental stage of the organism.
(6) Several characters can be screened in the same sample.

Probes

Finding RFLPs is dependent on the number of probes available. Probes are small segments of DNA or RNA of known base sequence, function and/or origin. Probes are used to detect the presence of nucleotide sequences similar to those of the probes in a living or non-living system. There is no immediate need to always find the gene(s) directly responsible for a character expressed at the phenotypic level. It should be sufficient for applied breeding strategies to detect characteristic RFLPs correlated to phenotypes. Correlations are of the utmost importance, particularly from the standpoint that the agriculturally important characters are not coded just by one gene but are of polygenic origin (Kuckuck *et al.*, 1993).

Both cDNA and random genomic libraries (see box) are used as sources of probes for RFLP mapping. Most RFLP laboratories use probes radioactively labelled with p^{32}. Non-radioactively labelled probes can also be used in RFLP analysis and are desirable for safety purpose and in situations where isotopes are difficult to obtain or to use due to hazardous radioactive effects (Brar and Dhaliwal, 1997). Most RFLP maps have been constructed using F_2 or backcross populations.

Disadvantages of RFLP

Conventional RFLP analysis has some disadvantages :

(1) RFLP requires large amount of DNA.
(2) In RFLP analysis, the Southern blotting and hybridization procedure need radioactive isotopes and autoradiography. These factors make RFLP analysis relatively slow and expensive.
(3) RFLP analysis requires considerable time.

RFLP Map Development in Rice

A molecular map of rice consisting of over 600 RFLP markers has been developed at Cornell University, USA (McCouch, 1993) and made available to rice researchers throughout the world. The existing map is based on an interspecific backcross population derived from a cross between *Oryza sativa* (indica) and *O. longistaminata*, developed in the Cote d'Ivoire, Africa. The high level of genetic diversity inherent in this cross makes it useful for genetic mapping because of the high level of polymorphism detectable at the molecular level. Several libraries of clones have been used to construct the map.

Box - 2

cDNA Library is a population of bacterial transformants in which each mRNA is represented as its cDNA insertion in a plasmid or a phage vector. It lacks introns.

Genomic library is a population of bacterial transformants in which each genomic DNA fragment is present as insertion in a plasmid or a phage vector. It contains both introns and exons.

The first published RFLP map of rice was based on random genomic clones derived from a Pst1 library (McCouch *et al.*, 1988). To enrich the genetic map for single copy sequences (which are most useful for mapping), several cDNA libraries were developed from a variety of plant species. The rice map currently includes sequences from both a genomic and a cDNA library of rice, cDNA clones derived from oat and barley, and genomic clones from maize.

The RFLP map of rice is one of the most densely saturated maps of any monocot species. Rice has the smallest genome of any monocot species known (ca. 5×10^8 base pairs) (Arumuganathan and Earle, 1991). Rice is rapidly emerging as the model monocot system for map-based gene cloning due to availability of several well characterised genetic mutants of rice. Rice genome has a total genetic distance of approximately 1800 cM (Saito *et al.*, 1991) and 1 cM is equal to approximately 250 - 300 kb. Therefore, the current level of saturation provided by the existing rice map is approximately one marker every 750 - 900 kb.

Efforts are being made to further saturate the genetic map of rice by integrating the map developed at Cornell University and that developed independently by researchers at Tsukuba City of Japan (Saito *et al.*, 1991). In addition, comparative mapping efforts in rice and maize offer a new opportunity for increasing the reservoir of available markers for rice (McCouch, 1993). These maps together provide rice researchers with 1000 RFLP markers mapped onto the 12 chromosomes of rice. Recently, Brar and Dhaliwal (1997) have reported that 2300 DNA markers have been mapped in rice genome.

Molecular markers are quite powerful to saturate the genetic map as compared with the classical map comprising of morphological markers. The classical map comprising of morphological and isozyme markers has less than 200 markers as compared with the molecular map (comprising RFLPs and other DNA markers) having at least 2300 DNA markers.

RFLP Gene Tagging and Marker-Assisted Selection in Rice

Gene tagging is the process of locating genes of interest via linkage to molecular markers. A molecular marker very closely linked to a gene can act as a "tag" which can be used for indirect selection of the gene in breeding programme. When RFLP markers are used in locating genes via linkage, the technique is called RFLP gene tagging.

In modern plant breeding the process of selection for genetic improvement is exercised on the basis of molecular markers. This selection is referred to as marker-assisted or marker-aided selection (MAS). In MAS, individuals carrying target genes are selected in a segregating population based on patterns of tightly linked markers rather than on their phenotypes.

In practice classical 'phenotypes' and molecular marker 'genotypes' are evaluated on the same individuals or related progenies. The data are analysed to determine if any of the RFLP markers co-segregates with the target phenotype. Once such co-segregation or correlation is established and the genes of interest are located, thus the RFLP markers can be used to determine whether the genes are present or absent in an individual or in a population without the evaluation of the phenotype. In this way the genotypes of interest can be identified at any stage of the plant in various environments without waiting for the phenotype to express at a specific developmental stage. This further enables evaluation of multiple characters simultaneously.

RFLP gene-tagging work in rice was directed towards several characters :

(1) **Disease resistance** : Earliest gene-tagging work in rice focused on finding markers closely linked to single genes for disease and insect resistance, such as blast resistance, bacterial blight

resistance and the white-backed plant hopper resistance (See Table 15.1). At present, tight linkage of several resistance genes with RFLP markers has been established due to the untiring efforts of several groups of dedicated workers.

Using RFLP gene-tagging technique, at least 8 genes conferring resistance for bacterial blight (BB) have been tagged. Efforts are being made at IRRI (Philippines) to combine several genes for bacterial blight resistance in a single variety which could not be done on the basis of phenotypic selection only. Without inoculating rice plants with *Xanthomonas oryzae* pv. *oryzae* (the bacterium causing bacterial blight disease in rice), it has been possible to select for a dominant gene and a recessive gene simultaneously conferring BB resistance in one generation.

Table 15.1: Some examples of useful genes tagged with molecular markers in rice

Gene Symbol	Character	Chromosome No.	Donor Variety	Linked Marker
Pi - 1 (t)	Blast resistance	11	LAC 23	Npb 181
Pi - 2 (t)	Blast resistance	6	5173	RG 64
Pi - 4 (t)	Blast resistance	12	Tetep	RG 869
	Blast resistance	12	—	RZ 536
Pi - z^5	Blast resistance	6	—	RG 64
	Blast resistance	6	—	RG 612
Pi - ta	Blast resistance	12	—	RZ 397
	Blast resistance	12	—	RG 241
Pi - 5 (t)	Blast resistance	4	Moroberekan	RG 498
Pi - 6	Blast resistance	12	—	RG 869

Table 15.1: *(Contd...)*

Gene Symbol	Character	Chromosome No.	Donor Variety	Linked Marker
Pi - 7	Blast resistance	11	—	RG 103
Pi - 9	Blast resistance	6	—	RG 16
Pi - 10	Blast resistance	5	—	RRF 6, RRH 18
Pi - 11	Blast resistance	8	—	BP 127
				RZ 617
Pi - ? (t)	Blast resistance	4	IRAT 13	—
Pi - 12	Blast resistance	12	—	RG 869
Pi - b	Blast resistance	2	—	RZ 123
Xa -1	Bacterial blight	4	Kogyoku	Npb 235
	(BB) resistance			Npb 197
Xa - 2	BB resistance	4		Npb 235
				Npb 197
Xa - 3	BB resistance	11	Chogoku 45	Npb 181
				Npb 78
Xa - 4	BB resistance	11	IR 20	Npb 181
				Npb 78
Xa - 5	BB resistance	5	IR 1545-339	RG 556
Xa - 10	BB resistance	11	CAS 209	—

Table 15.1: *(Contd...)*

Gene Symbol	Character	Chromosome No.	Donor Variety	Linked Marker
Xa - 13	BB resistance	8	—	CDO 116
Xa - 21	BB resistance	11	*O. longistaminata*	pTA 818
				pTA 248
				RG 103
RTSV	Rice tungro spherical virus resistance	4	—	RZ 262
Hbv	Hojablanca resistance	12	Fanny	—
Gm-2	Gall midge resistance	4	Siam 29	RG 229
				RG 476
GM	Gall midge resistance	4	—	MG 1400
Bph- 1	Brown planthopper resistance	12	—	XNpb 248
Bph-1G	Brown planthopper resistance	12	—	RG 457
ef	Early flowering	10	—	CDO 98
fgr	Fragrance	8	—	RG 28
Wph-1	Whitebacked planthopper resistance	7	—	N 22
Glh	Green leaf hopper	4	—	RZ 262
Rf	Fertility restorer	10	—	RG 561, CDO 94
Rf-2	Fertility restorer	1	—	RZ 382, RG 435

Table 15.1: *(Contd...)*

Gene Symbol	Character	Chromosome No.	Donor Variety	Linked Marker
Rf-3	Fertility restorer	1	—	RG 532
S-5	Wide compatibility	6	—	RG 138
		6	—	RG 213
sd-1	Semi dwarf	1	—	RG 109
sco(d)	Semi dwarf	6	—	RZ 182
PMS-1	Photoperiod sensitive male sterility	7	—	RZ 477
PMS-2	Photoperiod sensitive male sterility	3	—	RG 191
tms-1	Thermosensitive male sterility	8	—	—
tms-3	Thermosensitive male sterility	8	—	OPAC 3-640
Se-1	Photoperiod sensitivity	6	—	RG 64
Se-3	Photoperiod sensitivity	6	—	A 18
Sub-f	Submergence tolerance	3	—	RZ 698

The dominant resistance gene Xa-21, which confers BB resistance to all 6 internationally recognised races of Xoo in the Philippines, has been mapped to the top of chromosome number 11 of rice. This gene was introgressed into cultivated rice from the wild species, *Oryza longistaminata* (Khush *et al.*, 1991). The recessive gene Xa-5, which confers resistance to 4 of the 6 Philippine races of Xoo, has been mapped to the top of chromosome 5 (McCouch *et al.*, 1991). Both the genes are tightly linked to 3 RFLP markers in the respective regions.

At IRRI, workers are endeavouring to incorporate other BB resistance genes into a "superdonor" line and to convert tightly linked markers to primers appropriate for PCR-based selection (McCouch 1993). The superdonor line will provide the breeders with the opportunity:

(1) to select multiple genes conferring resistance to BB disease in a single cross, and
(2) to test how various combinations of BB resistance genes behave in response to disease pressure.

In order to develop a rice variety with durable resistance to the blast fungus (*Pyricularia oryzae*), attempts have been made at IRRI to combine multiple genes conferring both qualitative and quantitative resistance. Qualitative resistance genes are short-lived and race-specific providing complete resistance whereas quantitative resistance genes are race non-specific and confer partial resistance. Single, race-specific genes for resistance to blast have proved short-lived in the field because they impose an intense selection pressure on the pathogen. The pathogen tries hard to overcome the selection pressure possibly by mutation. Multiple genes conferring partial blast resistance are extremely difficult to combine in the same plant using conventional approaches because they produce confusing phenotypes when combined, thereby making selection of the plant difficult. But molecular markers provide an efficient means of identifying genotypes containing multiple resistance loci. Both qualitative race-specific and quantitative race non-specific genes for blast resistance have been tagged (Yu *et al.*, 1991; Wang *et al.*, 1992).

(2) **Abiotic stress tolerance** : Genes conferring tolerance or resistance to stresses such as water stress, saline stress and mineral deficiency or toxicity stress are being tagged at IRRI. There are some hurdles in tagging genes for abiotic stress tolerance :

(i) Screening techniques for identifying stress tolerant

genotypes are not highly developed.

(ii) Mechanisms of tolerance or resistance are poorly understood.

(iii) In some cases, tolerance or resistance is not maintained in the absence of stress.

(3) **Quality and adaptability :** Traditional rice varieties are of excellent grain quality and well adapted to the ecological niche where they have been grown by farmers for a long time. But yields are often very low. Efforts have been made to transfer the genes for quality and adaptability from traditional rice varieties into improved, high yielding varieties by identifying genetic loci that contribute these phenotypes and using RFLP markers to assist their selection.

(4) **Hybrid rice :** Single genes governing thermosensitive male sterility and fertility restoration have been tagged in rice. This achievement has ushered a new chapter in hybrid rice production. Besides this, the RFLP marker technology has been utilised not only in understanding the phenomena of heterosis and transgressive variation in rice but also in maximizing the efficiency of crossing schemes for developing high performance varieties (McCouch, 1993).

RFLP Markers in Utilizing Cryptic Variation in Rice

Wild germplasms abound in genetic variation. The tremendous range of naturally occurring genetic variation makes rice one of the most widely adapted crop species known. Cryptic genetic variation often goes undiscovered by the breeders because target characters are not necessarily identified in the germplasms. RFLP markers have been utilised to tap hidden source of genetic variation in rice. Efforts have been made to tag alleles in the wild germplasm that contribute to transgressive variation in wide crosses. Once identified, these novel alleles could be transferred to the improved and high yielding backgrounds. RFLP markers have been used to tag a bacterial blight resistance gene derived from the wildspecies *Oryza australiensis* in the early generation backcross progeny of

O. sativa x *O. australiensis*. In such interspecific crosses, RFLP markers have several advantages :

(1) RFLP markers offer a way to select for individuals carrying the smallest segment of "Wild" DNA in the vicinity of the gene(s) of interest

(2) RFLP markers help in rapidly returning to parental types in a backcross programme because the level of polymorphism is very high in most interspecific crosses and the markers help in identifying parental types in this effort.

In order to facilitate further mapping of random markers and gene of interest, an interspecific backcross population derived from *O. sativa* and *O. longistaminata* is maintained vegetatively at IRRI, Philippines and at Cornell, University of Georgia, USA.

Randomly Amplified Polymorphic DNA (RAPD)

The generation of RAPDs involves the use of single short random oligonucleotides. Oligonucleotides are short chains of nucleotides. When these random oligonucleotides are mixed with genomic DNA and thermostable DNA polymerase (enzyme that catalyses the systhesis of DNA) and subjected to PCR (Polymerase chain reaction), then the amplification of several genomic DNA fragments starts. Particular fragments of genomic DNA are amplified to large quantities which help in their detection. The amplified portions of DNA are defined by short (16 – 20 base), synthetic DNA molecules (oligonucleotides/primers) that match the ends of genomic DNA to be amplified.

Williams *et al.*(1990) developed RAPD procedure, in which 10-base primers (oligonucleotides) of arbitrary sequence are used to amplify different DNA fragments using genomic DNA as template. The basic principle behind RAPD is that by virtue of the short primer sequence, there is a high probability that two priming sites will occur in the genome in inverted orientation and close to each other. The intervening region between two priming sites is amplified and results in different sizes of DNA fragments. These fragments can be separated by gel electrophoresis. The amplified DNA fragments behave as simple Mendelian markers. The DNA amplification with random primers exposes polymorphisms distributed throughout the genome. Polymorphic bands are noted in the gel electrophoresis and the polymorphisms can be mapped in a segregating population.

Advantages

(1) In RAPD analysis, prior knowledge of DNA sequence is not necessary to design the primers. So, abundant genetic markers can be generated.
(2) RAPD is actually a PCR-based marker. RAPDs can be utilised in much the same way that RFLPs are used.
(3) RAPDs can be digested with a series of restriction enzymes and the size of restriction fragments can be compared by gel electrophoresis.

Disadvantages

(1) RAPDs are dominant markers (unlike RFLPs which are codominant) thus preventing accurate detection of heterozygotes (Williams *et al.*, 1990).
(2) RAPDs are not repeatable and hence may not amplify the same genomic DNA fragments.
(3) RAPDs are not stable.

Polymerase Chain Reaction (PCR) Based Markers

Polymerase chain reaction (PCR) is a recently developed technique in the field of molecular biology. PCR was invented by Kary B. Mullis in 1987 (Mullis and Faloona, 1987) and first used for an application by the department of Genetics of the Cetus Corporation of USA. PCR is an *in vitro* method for the enzymatic synthesis of specific DNA sequence through amplification by a factor of up to 10^6 fold.

be used to study the extent of genetic variation among different individuals at the molecular level directly.

Chloroplast DNA (cpDNA) and mitochondrial DNA (mt DNA) are smaller in size than the nuclear DNA. Usually the cpDNA produces nearly 40 fragments after digestion with a typical restriction enzyme like EcoRI. RFLP analyses of cpDNA and mtDNA are very useful in studying the phylogeny or systematic relationship. In plant breeding the RFLP analyses of cpDNA and mtDNA are of limited importance because most of the genes governing characters of agronomic utility and productivity are located in the nuclear DNA. Hence RFLP analysis of nuclear DNA finds greater applicability in plant breeding. Due to large size of nuclear DNA, the restriction enzyme produces a very large number of fragments in a continuous range of sizes. When the digested nuclear DNA is subjected to agarose gel electrophoresis and stained with ethidium bromide, no distinct fragments can be seen in the gel. So many fragments are produced that the nuclear DNA appears as a continuous smear on the gel.

Agarose Gel Electrophoresis (AGE)

For separation, identification and purification of DNA fragments, agarose gel electrophoresis is used. Since DNA has a large number of phosphate groups which are negatively charged at neutral p^{H}, it will migrate toward the anode (+ve electrode). The basic phosphodiester backbone of all DNAs is the same, so they have a uniform charge density. Therefore, if the DNA is placed in a porous medium such as agarose or polyacrylamide, it will migrate toward the anode at a particular migration rate. Several factors influence the migration rate of DNA fragments through the agarose gels (Fig. 16.1). This technique is commonly used to separate DNA fragments of different sizes produced by digestion of DNA with restriction enzymes. The agarose gels are commonly run in horizontal position.

Factors Influencing Migration Rate

The migration rate of DNA fragments through agarose gels depends

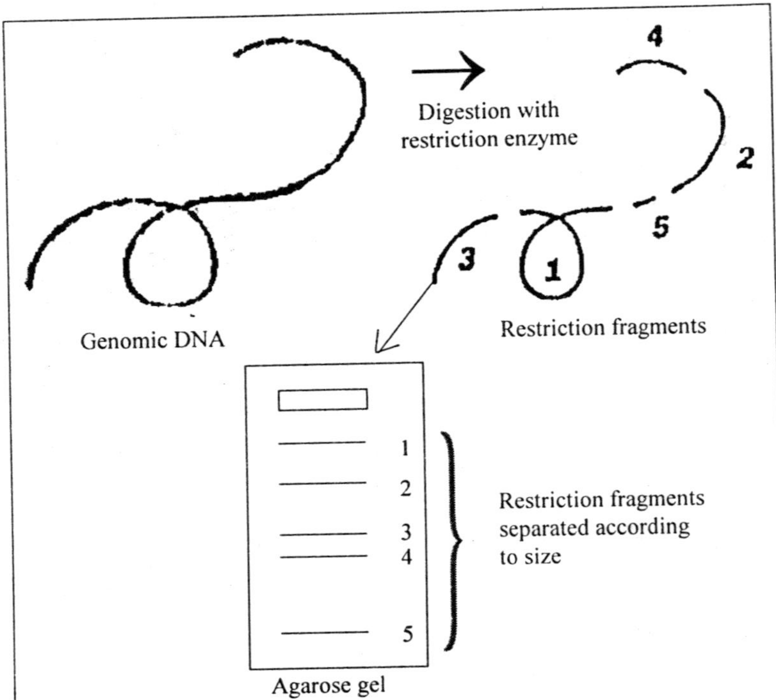

Fig. 16.1: Agarose Electrophoresis

on several factors (Maniatis *et al.*,1982), which are described below.

(1) **Molecular weight of the DNA** : Linear, duplex DNA molecules migrate through agarose gels at rates that are inversely proportional to the $\log_{10}$ of their molecular weights.

(2) **Agarose concentration** : A particular DNA fragment moves at different rates through gels containing different concentrations of agarose. Using gels of different agarose concentration, it is possible to separate a wide size-range of DNA molecules.

(3) **Current flow** : At low voltages, the migration rate of linear DNA fragments is proportional to the voltage applied. But at high voltages, the migration rate of high

molecular DNA fragments increases differentially. To obtain maximum resolution of DNA fragments, gels should not be run above 5 volts/cm.

Agarose concentration (%)	Efficient range of separation of Linear DNA molecules (kb)
0.3	60 - 5
0.6	20 - 1
0.7	10 - 0.8
0.9	7 - 0.5
1.2	6 - 0.4
1.5	4 - 0.2
2.0	3 - 0.1

(4) **DNA conformation :** DNA molecules having the same molecular weight may exist in three different conformations viz. closed circular (form - I), nicked circular (form - II) and linear (form - III). The migration rate of different conformations through agarose gels varies depending on the agarose concentration, the strength of the current, the ionic strength of the buffer and the density of superhelical twists in the form-I DNA. In some conditions, the closed circular DNA moves faster than the linear DNA but in other conditions the pattern is reversed.

Base composition of DNA and temperature (between 4°C and 30°C) do not significantly affect the migration rate of DNA fragments through agarose gels.

Polyacrylamide Gel Electrophoresis (PAGE)

Polyacrylamide gels are used to analyse the DNA fragments less than 1 kb (kilobase) in length. The gels are invariably run in the vertical position. The gels are prepared or cast in different concentrations ranging from 3.5% to 20%, depending on the sizes of DNA fragments.

Polyacrylamide is toxic and is absorbed rapidly through the skin. So the user must wear gloves in handling the polyacrylamide gel as a safety measure.

Acrylamide Concentration(%)	Effective range of separation (nucleotides)
3.5	100 - 1000
5.0	80 - 500
8.0	60 - 400
12.0	40 - 200
20.0	10 - 100

Separation of Restriction Fragments

In RFLP analysis, any plant tissue or organ (seed, stem, leaf, root or flower etc.) can be used for isolation of DNA. At first, the plant organ is cut into small pieces and ground to a fine powder in presence of liquid nitrogen. The fine powder is mixed with several solutions in required quantities as per the standard procedure and thus subjected to centrifugation to precipitate the DNA. It involves a series of steps. The isolated DNA is dissolved in a suitable buffer and then purified. The methods for isolation of plant DNA and its purification have been discussed in detail by Sadasivam and Manikam (1992).

The purified DNA is mixed with a suitable restriction enzyme to cleave the DNA at specific restriction sites. After digestion with the restriction enzyme, the large DNA molecules are reduced to a series of fragments varying in size. The restriction fragments are separated by agarose gel electrophoresis (AGE) or polyacrylamide gel electrophoresis (PAGE).

The restriction fragments produce a pattern in the gel and this pattern is specific for the particular DNA and restriction enzyme combination. Genetically different varieties or lines of a crop species produce different RFLP patterns. These differences in the RFLPs form the basis in determining the extent of genetic variation among the varieties or lines.

Production of Cloned Probes

Probes are labelled, single stranded DNA or RNA molecules of

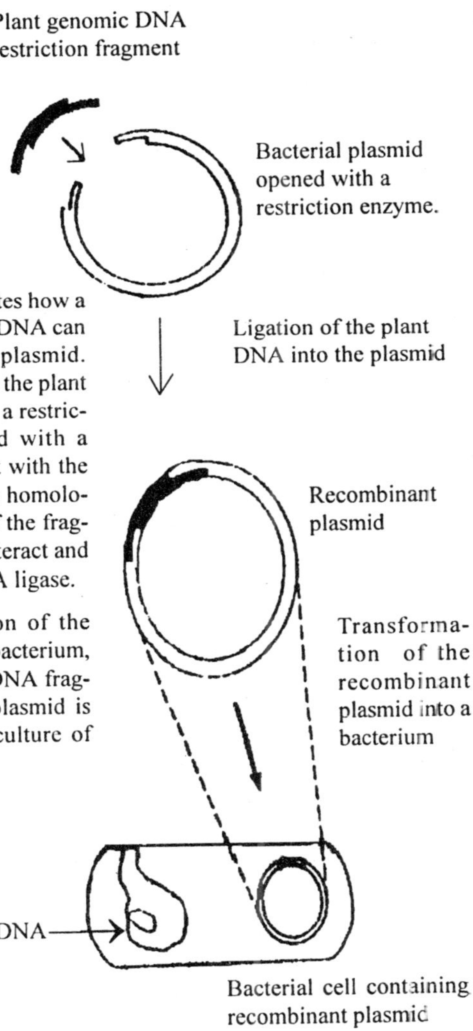

This diagram illustrates how a restriction fragment of plant DNA can be cloned into a bacterial plasmid. DNA would be isolated from the plant of interest and digested with a restriction enzyme. When mixed with a plasmid, which has been cut with the same restriction enzyme, the homologous single stranded ends of the fragment and the plasmid can interact and be covalently joined by DNA ligase.

After transformation of the recombinant plasmid into a bacterium, large amounts of the plant DNA fragment are produced as the plasmid is replicated in the growing culture of bacteria.

Fig. 16.2: Cloning of Plant DNA

size about 2-5 kb. They are used to detect individual DNA fragment of interest in a complex mixture of different fragments. Probe hybridises with the DNA fragment of interest and it helps in detection of the latter.

RFLPs of nuclear DNA cannot be directly observed or studied, so small pieces of chromosomal DNA are used as probes. Restriction fragments generated by restriction enzyme can be used as probes but selection of the desired restriction fragment (double stranded DNA) in a pure form from the mixture of fragments is needed. It is difficult to select and isolate a restriction fragment but fortunately it is not necessary to do this. This can be easily done by gene cloning technology with *E. coli* bacterium as a host. In this technique, individual restriction fragments are ligated (joined) into bacterial plasmids (Fig. 16.2). A plasmid is a self-replicating circular DNA molecule which can incorporate foreign DNA into it and occurs free-floating in bacterial cytoplasm. The plasmid carrying the unknown individual restriction fragment (called recombinant plasmid) is transformed (introduced) into a cell of *E. coli* bacteria. The transformed bacterium grows in the culture medium and multiplies producing a large number recombinant plasmids. It gives a large supply of a particular restriction fragment which is suitable for use as a hybridization probe. Since the probe is produced by cloning, it is called a cloned probe. The desired probe can be selected from the particular bacterial strain using specific method. Each bacterial strain contains only a specific restriction fragment carrying plasmid.

Detection of RFLPs in Nuclear DNA using Probes

DNA from the plants to be compared for RFLP differences is isolated separately, digested with a restriction enzyme and then separated (fractionated) on an agarose gel. The DNA from each plant will now be in the form of a large number of restriction fragments separated in the gel according to weight. To detect specific fragments by DNA-DNA hybridization, the cloned probe DNA and the DNA of each plant in the gel must be single stranded. Single stranded DNA are produced from double stranded DNA by the process of denaturation using either alkali treatment or high temperature treatment.

The DNA in the gel is soaked in NaOH solution (alkali) to denature the DNA. The single stranded DNA thus produced is transferred from the gel to a nylon membrane or filter by a process called "Southern" transfer (named after E.M. Southern who invented this technique) to facilitate hybridization. A nylon membrane cut to the same size of the gel is placed on the gel and as a result the denatured DNA (single stranded) is transferred to the nylon membrane. In this process the nylon membrane gets ready for hybridization with the cloned probes of the library.

The cloned probes are also double stranded. They are made single stranded. Thus under proper temperature and salt concentration, the single stranded probe will hybridize specifically to the single stranded DNA on the nylon membrane which are complementary to each other. However, in order to see whether the cloned probe has hybridized or not, it is necessary to label the probe in some way. One way is to label the probe radioactively with P^{32}. The labelled single stranded probe is allowed to hybridize with the complementary single stranded DNA on the nylon membrane and the excess unhybridized probe is washed off the nylon membrane. Then the nylon membrane is subjected to autoradiography in order to detect the restriction fragments to which the probe has hybridized. Since radioactive element is not ecofriendly, nowadays other nonradioactive substances are used for probe labelling.

RFLP Inheritance Pattern

RFLP markers strictly follow Mendel's laws of segregation and independent assortment. In classical genetics, morphological character with contrasting forms were chosen as genetic markers *e.g.* plant height (dwarf and tall), seed shape (round and wrinkled), seed colour etc.

In order to understand the inheritance of RFLP marker, we shall consider here a cross between two true-breeding (homozygous) parents differing for a conventional single gene

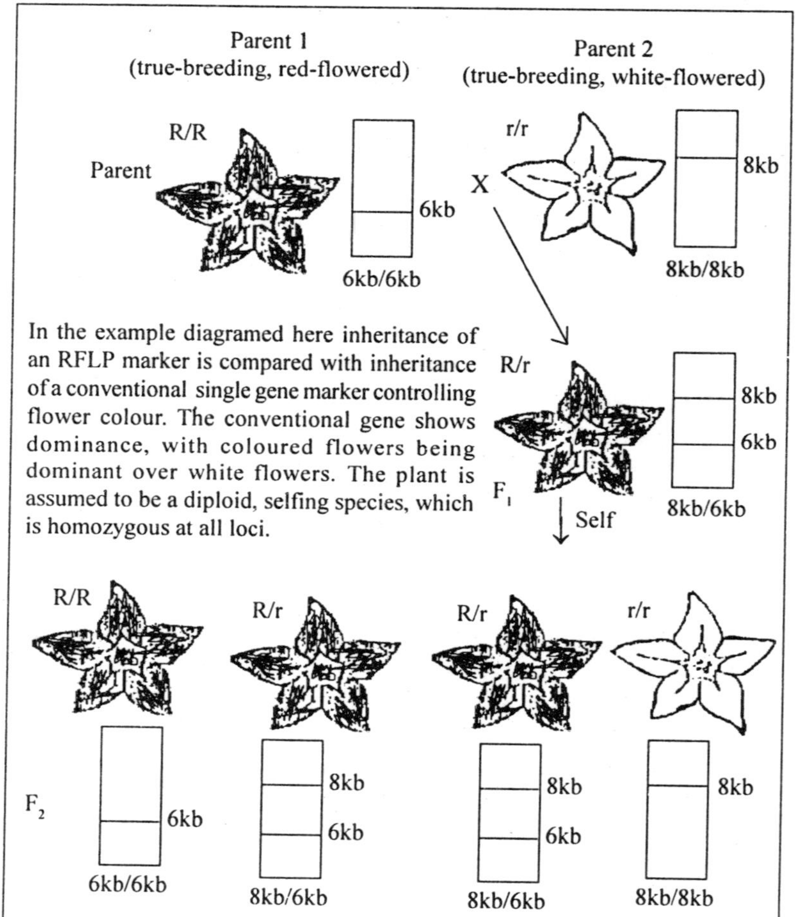

Fig. 16.3: RFLP Inheritance Patterns

marker controlling flower colour (Fig. 16.3). Parent 1 (R/R) is red-flowered which is caused by a 6 kb DNA fragment whereas parent 2 (r/r) is white-flowered caused by a 8 kb fragment. The F_1(R/r) will be red -flowered and both 6kb and 8kb fragments will appear in the gel following restriction digestion of its DNA. When the F_1 is selfed, it will produce three different F_2 genotypes *viz.* R/R (6 kb/6 kb), R/r (8 kb/6 kb) and r/r (8 kb/8 kb) in the ratio 1:2:1 but only two phenotypes (red and white flower) in the ratio 3:1 provided that R is completely dominant over r allele. From Fig. 16.3, it is evident that a RFLP marker (8 kb/6 kb) segregates and recombines exactly in the same way as a conventional gene marker (R/r). This means that segregation patterns of RFLP markers can be analysed by conventional Mendelian methods. Furthermore, maps of RFLP markers can be constructed in the same fashion as the maps of conventional markers.

Construction of RFLP Genetic Map

In the construction of a RFLP genetic map, a series of steps are involved. These are described below:

(1) **Selection of parents:** Genetically different plants must be chosen as parents. The parents should be genetically divergent enough to exhibit sufficient RFLP but not so far distant as to cause sterility of F_1. The parents are crossed to produce F_1 plants.

(2) **Production of mapping population:** The F_1 plants are selfed to produce the F_2 plants, which serve as the mapping population. A mapping population consisting of 50 F_2 plants is sufficient for construction of a fairly detailed map. Sometimes, instead of F_2 plants, the F_1 may be backcrossed to either of the two parents to develop the backcross progeny. These backcross progeny can also be used as a mapping population. However, it is better to have F_2 progeny as a mapping population than the backcross progeny because more information can be gained from the F_2 progeny.

(3) **Scoring RFLPs in the mapping population:** Once the mapping population is obtained, DNA is isolated from each individual plant of the population and subjected to gel electrophoresis. Each plant in the mapping population contains a unique array of parental chromosome segments. To derive a RFLP map it is necessary to determine which

parent's chromosome segments are present in each plant of the mapping population. It is further essential to keep each F_2 plant alive for extraction of DNA repeatedly. Each F_2 plant is unique for chromosome combination which will be destroyed by recombination if the F_2 plant is allowed to produce the F_3 progeny. Mapping thus proceeds by sequentially scoring RFLPs of the individual plants in the mapping population in the following procedure:

(A) **Screening for polymorphism** : Cloned probes are selected sequentially from the library and tested against the parents in order to determine which restriction enzymes will detect a polymorphism between the parents. The number of restriction enzymes will depend on the amount of variation present in the parent plants. If the parent plants are genetically close, a large number of restriction enzymes will be essential. Highly polymorphic plants, such as inbred lines of maize, will usually require only one or two enzymes to detect polymorphism with most probes. In rice, 11 enzymes were used and about 75% of the probes were polymorphic with these.

(B) **Scoring:** When a polymorphic probe / enzyme combination is detected, it can be scored in the mapping population. For this a series of agarose gels are prepared and then nylon membranes or filters are prepared from them for hybridization. These filters are called F_2 survey filters or mapping filters. There must be one filter set for each restriction enzyme and each filter set must contain digested DNA from each individual plant in the mapping population. For example, one set of mapping filters would contain DNA from each of the F_2 plants digested with EcoRI (restriction enzyme) whereas another set would contain DNA from each of the same F_2 plants digested with Pst I (restriction enzyme). If a polymorphism between the parent plants has been detected with a given probe/ enzyme combination, then that probe will be scored on the corresponding mapping filter. In a F_2 mapping population scored for a single RFLP, only three types of plants can be present—dominant homozygote, heterozygote and recessive homozygote. Each plant in the mapping population can be scored as one of the two possible homozygotes or as a heterozygote.

(C) **Linkage analysis :** Linkage analysis is based on the degree to which probes tend to cosegregate. The first probe scored in the mapping study will provide no information on linkage. If the second probe is linked to the first one, they will tend to co-segregate *i.e.* F_2 plants which are heterozygous for the first probe will also be heterozygous for the first probe. Similarly, plants homozygous for the allele found in one of the parents for the second probe will also be homozygous for the second probe. But in the absence of linkage between two probes, the distribution of homozygotes and heterozygotes for these two probes will be random. The randomness of segregation can be tested by simple statistical test such as a chi-square test.

The first few probes may not be linked and may not co-segregate because these probes will be randomly selected. As each probe is screened, it is tested for possible linkage to all other markers which have been mapped. In such analysis, a great amount of data is accumulated and a computer is used for data storage and analysis. When a new probe is scored and linkage is indicated with one or more markers, then the map distance is determined by computer analysis using an algorithm such as maximum likelihood. An algorithm is a procedural model for complicated calculation and it solves a problem in a few steps. Commonly used mapping functions are utilized to convert recombination data to map distances.

Advantages of RFLP Map

RFLP mapping project aims at the production of a saturated genetic map. A map with RFLP markers spaced every 10-20 centimorgans (cM) over all the chromosomes is called a saturated genetic map. On such a map, any conventional gene will be located within a few centimorgans of a RFLP marker. A conventional plant genetic map is about 1500 centimorgans, therefore 150 well-spaced RFLP markers would be sufficient for a saturated map. But in actual practice, probes are randomly selected and therefore it is necessary to map several hundred markers (not exactly 150) to achieve the desirable saturation.

Once a saturated or nearly saturated RFLP map is constructed, it can be used for locating conventional genes. For this it is essential to have a mapping population in which both the conventional gene and the RFLP markers are segregating. One can place the conventional gene on the map by determining which RFLP markers show co-segregation with the conventional marker.

Advantages of RFLP Markers

RFLP markers have several advantages over the conventional genetic markers and are described below:

(1) **Natural variation**: The base sequence of DNA varies in different organisms. RFLP markers are based on naturally occurring variation in DNA base sequence. Unlimited numbers of RFLP markers are available in virtually any plant species. Conventional genetic markers are based on contrasting phenotype of parents. One parent must have wild type allele whereas the other parent must have the mutant allele, mostly recessive, in homozygous condition in order to follow the allele through segregation for constructing a genetic map. Often the mutant allele is not detectable by its phenotypic expression. In such cases, mutant allele must be produced by mutagenesis which involves creation of variation. But RFLP marker utilizes abundant natural variation in DNA.

(2) **Phenotypic effect** : RFLP markers reflect the natural variation present in the base sequence of the DNA, including introns and exons. RFLP markers are unlimited

and most of the markers will have no phenotypic effect as they represent the introns. Introns are intervening sequences in DNA and do not usually code for proteins. Exons are coding sequences of DNA.

In conventional genetic mapping, only a few externally visible phenotypic markers which are encoded by the exons only are mapped, keeping aside the variation present in the introns which have no phenotypic effect in the organism. Moreover, it is very time consuming to bring all the mutant traits from different plants into a single plant through crossing and then to observe the segregation pattern of the mutant phenotypic markers in conventional mapping. A single plant contains large amount of RFLP variation. It is, therefore, possible to construct a detailed RFLP map using a single cross (hybrid) and a single mapping population (one F_2 population) by a single research group. But in conventional mapping, hundreds of crosses and mapping populations are required along with the co-operation of several research groups.

(3) **Developmental stage and plant organ**: In RFLP mapping, DNA isolation from any plant organ like leaf, stem, root, flower, seed etc. at any developmental stage can be used for RFLP map construction.

But in conventional mapping, it is necessary to wait for a particular developmental stage for scoring the morphological marker like flower colour, insect resistance etc. For example, in order to study the segregation pattern of flower colour in F_2 generation, one has to wait till flowering stage of the plant. To score for an insect resistance gene, it is often necessary to build cages around plants and introduce insects. Such study is time consuming and labour intensive.

But RFLP mapping overcomes this problem. Any developmental stage and any plant organ from which a small amount of DNA can be isolated are suitable for RFLP mapping.

(4) **Environmental effect** : Conventional morphological markers depend on gene expression for scoring. The DNA sequence representing the gene must be transcribed to produce mRNA which is then translated to produce a specific protein. That protein may either function as an enzyme or bind to some cell component and finally produces the phenotype. Expression of the DNA representing the morphological marker is influenced by both internal and external environment.

Since scoring of RFLP markers does not depend on gene expression, RFLP markers are not affected by environmental factors. RFLP markers simply represent the presence or absence of characteristic base sequence.

(5) **Dominance and Epistasis** : Conventional morphological markers depend on gene expression for scoring. So the phenotypic effect is often modified by intra-allelic (dominance) or inter-allelic (epistasis) interaction.
For example, it is difficult to score a recessive gene (a) unless the gene occurs in a recessive homozygote (aa). The recessive gene cannot be scored in a dominant heterozygote (Aa) because the dominant allele (A) does not permit expression of the recessive allele. Moreover, phenotypic effects of some genes are modified by the alleles of other genes which may be in the same chromosome or a different chromosome. Scoring conventional morphological markers becomes complicated due to dominance and epistasis.
RFLP markers are co-dominant. Both the dominant and the recessive alleles present in the heterozygote can be scored in RFLP analysis.

Uses of RFLP in Plant Breeding

A RFLP map is not useful by itself in plant breeding but it becomes useful when it is used in association with analysis of conventional markers. Sometimes in conventional plant breeding, direct selection of a conventional marker/gene is difficult, expensive and time-consuming. In such cases, indirect selection for the conventional gene is done by selecting the RFLP marker with which the conventional gene is linked or correlated. Such selection is called "RFLP marker aided selection".

Indirect selection for a RFLP marker, rather than selection for the conventional gene, may have advantages in several plant breeding situations:

(1) **Selection for recessive gene** : Conventional breeding approach of transferring a recessive gene from one cultivar to a desirable cultivar requires alternate backcross and selfing. The recessive gene comes to homozygous condition through selfing of the heterozygote and then the recessive gene is selected by selecting the recessive homozygote with a specific phenotype. This is a time-consuming method and requires several generations.

If the recessive gene is linked to a RFLP marker, selection for the RFLP marker will easily select the recessive gene indirectly. This will save time and will expedite the breeding programme.

(2) **Simultaneous selection for multiple genes** : In resistance breeding against insects and diseases, it is often desirable to include more than one resistance gene to a particular insect or pathogen in the newly developed variety. If one resistance gene fails, then other resistance genes will save the crop against particular insect or pathogen.

In conventional breeding, selection for more than one resistance gene is difficult because the phenotype will be same whether one, two or more resistance genes are present. In such cases, progeny testing also is not very useful.

When linkage of each resistance gene to RFLP marker is established, then each resistance gene can be selected. Individual plants can be identified and easily selected for the presence of one, two, three or more resistance genes without progeny testing.

(3) **Selection for QTLs :** Most of the traits of agronomic importance like yield, quality etc. are governed by a large number of genes each with a small positive or negative effect. These genes are called polygenes or QTLs and scattered over the chromosome complement. The characters governed by these genes are called quantitative or polygenic characters. QTLs do not exhibit clear dominance and the phenotype is largely influenced by environment.

QTLs are studied with the help of biometrical methods which require extensive testing over years and locations. Before the advent of RFLP techniques, fundamental advances in biometrical genetics seemed unlikely, particularly with the location of QTLs on chromosomes.

With the help of RFLP analysis, it is now possible to establish the correlation between a particular polygenic trait and specific chromosome segment marked by RFLP. If correlations are established, then the chromosome segment must contain one or more genes determining the polygenic trait. The polygenic trait is scored following conventional method and the RFLP markers are scored easily.

Several RFLP markers involved in a particular polygenic trait can be scored in individual plants. Specific RFLP markers can be selected from a population of plants and incorporated into a single plant with high efficiency to produce superior plant.

REFERENCES

Kochert , G. 1990 . Introduction to RFLP mapping and plant breeding applications. Rockefeller Foundation International Program on Rice Biotechnology, Univ. of Georgia, USA.

Paterson, A.H. , Tanksley, S.D. and Sorrells, M.E. 1991. DNA markers in plant improvement. *Advances in Agronomy,* 46: 39 - 90.

17

Quantitative Trait Loci (QTLs)

Introduction

Plant characters of economic importance such as yield, quality, grain shape or size, disease and pest resistance, abiotic stress tolerance etc. are influenced by many genes. These genes have small, similar (positive or negative) and cumulative effects on the particular phenotype. These genes are called polygenes or minor genes or quantitative trait loci (QTLs). The characters governed by them are called quantitative or polygenic characters. The QTLs also follow Mendelian inheritance like major genes. They undergo segregation and recombination but are largely influenced by the environments.

The subject of quantitative genetics or biometrical genetics is based on the knowledge of polygenes as a whole but without the actual knowledge of the individual genes themselves. We do not know:

(1) How many genes govern a quantitative character ?
(2) What are the individual effects of these genes ?
(3) Where are these genes located on the chromosomes ? Or How are these genes distributed throughout the genome ?
(4) Are there "hot-spots" on particular chromosomes for the location of these genes governing a particular character ?

All these questions could be answered by mapping the QTLs governing a specific quantitative character.

QTLs —An Old Concept

The concept of QTLs is not new. It existed as early 1900 when Mendel's laws were rediscovered by three famous geneticists. Actually the molecular dissection of quantitative traits began before the demonstration that DNA is the genetic material.

The first account of linkage of a quantitative trait to a major gene locus was provided by Sax (1923). He studied the inheritance of seed weight and seed colour in the F_2 generation of a cross between two lines of the bean, *Phaseolus vulgaris*. Seed weight is a quantitative trait which shows continuous variation whereas seed colour is a qualitative character governed by a single gene (P/p) showing discrete variation. A qualitative character can be easily classified into distinct classes depending on the expression of the character.

Among the F_2 population, Sax (1923) observed that the individuals which were PP (dominant homozygote) for colour had heavier seeds than the individuals possessing pp (recessive homozygote) colour with lighter seeds. The heterozygotes (Pp) were exactly intermediate for seed weight. He concluded that :

(1) Factor differences for seed weight are linked with factors that determine the colour of the seed. Size differences may be effected by the independent action of factors in different linkage groups.
(2) These factors, when combined, have a cumulative effect.
(3) The size factors in different chromosomes may not be equal in their effect.

Thoday (1961) was of the opinion that the practical limitations in identifying the QTLs governing a quantitative character were the availability of very few genetic markers and the time that could be given to that work. At that time, however it was not possible

(1) to resolve the effect of individual QTLs,
(2) to tag individual QTLs to the easily scorable markers, and

(3) to practise directed selection of individual QTLs.

Advantages of Locating QTLs

(1) Locating QTLs of a quantitative character provides a means of answering some fundamental questions about the genetic control of the character *e.g.* number of genes, location of these genes in chromosomes etc. Such information will be of great help for formulating a successful breeding programme for the genetic improvement of the trait.

(2) Exact location of the genes on particular chromosomes helps to distinguish between the effects of linkage and those of pleiotropy on the characters. Pleiotropy refers to the situation when the same gene(s) controls two or more traits. Linkage is the tendency of two or more genes to be inherited together due to their presence on the same chromosome.

Knowing the occurrence of linkage or pleiotropy as a genetic cause is of great concern to the breeder because linkage between a desirable trait and an undesirable trait could be broken by recombination. On the other hand, if the two characters are due to pleiotropy, it leaves no opportunity to the breeder for rejecting the undesirable trait in his endeavour for improving the desirable trait.

(3) QTL analysis provides an opportunity to study not only additive and dominance effects of individual loci but also the epistasis (interaction between different loci).

(4) Location of QTLs improves the selection efficiency particularly when the QTLs of interest are linked to easily recognisable genes (*e.g.* qualitative character or marker loci). Such marker loci identify the presence of particular QTLs and allow marker assisted selection (MAS) of QTLS.

(5) The marker loci improve the selection efficiency for QTLs by selecting individuals at a juvenile stage when the QTLs are yet to express at phenotypic level.

(6) QTL analysis of economically important trait opens up the

possibilities of gene cloning and transformation of useful plants for higher yield and better quality.

(7) QTL analysis in diverse species helps to determine colinearity among genomes.

Linkage between QTL and Genetic Marker

For simply inherited traits, a difference between parents in one or two genes explains virtually 100% of the differences among the progeny – such a trait is itself a genetic marker. But most traits of agricultural importance are influenced by many genes. Further these traits tend to be quantitative, in that the progeny do not fall into discrete classes but show a continuous range of phenotypes between parental phenotypes and occasionally beyond the parental phenotypes (transgression).

Experiments with beans (Johanssen, 1909), wheat (Nilsson-Ehle, 1909) and tobacco (East, 1915) showed that such continuous variation could be explained by polygenes or QTLs (Geldermann, 1975). The QTLs exhibit the fundamental Mendelian properties of segregation and recombination. The QTLs and the genes affecting simply inherited traits differ little. Consequently it seemed likely that one might identify linkage between QTLs and genetic markers (genes affecting simply inherited traits) and thus determine the locations of individual QTLs.

Linkage between a genetic marker and a QTL was first demonstrated by Sax (1923) in *Phaseolus vulgaris*. Later on, in several other species like tobacco, maize, pea, bean, barley and tomato, associations between QTLs and visible genetic markers were observed. But many of these early studies had rather few genetic markers to work with. This was a serious limitation because it was possible to show a QTL in the vicinity of a particular marker but due to insufficient genetic markers it was impossible to pinpoint the location of the QTL.

Later on, genetic maps with more molecular markers (saturated genetic maps) were developed in several species and it became possible to more precisely estimate the location of a QTL by studying several markers along a chromosome. The limitation was, in fact, remedied by the construction of complete RFLP linkage maps, thus permitting systematic searches of an entire genome for QTLs influencing a trait (Paterson *et al.*, 1988). Further, new algorithms for QTL mapping minimised the number of individuals and number of genetic markers needed to map QTLs (Lander and Botstein, 1989). Algorithm is a procedural model for complicated calculations.

Description of Individual QTL

Using DNA markers, QTLs can be described by their chromosome location, dosage effect, phenotypic effect(s) and sensitivity to environment (Paterson *et al.*, 1991b). Such a description has long been possible for genes affecting simply inherited traits. But it has become possible to obtain this information for individual QTLs only after the availability of high density genetic maps (*i.e.* genetic map with more markers).

Chromosomal Location of QTL

The chromosomal location of a QTL can be determined by identifying nearby genetic markers which are usually cotransmitted with the QTL from parent to progeny. The same principle also applies to chromosomal location of a gene influencing a simply inherited trait (*i.e.* a gene with large effect). Using even a single genetic marker it is possible to determine that a QTL lies somewhere in the vicinity of the genetic marker. But several genetic markers along the chromosome are required to determine the exact location of the QTL. With a complete map of genetic markers, powerful statistical methods (Lander and Botstein, 1989) are employed to determine the likelihood intervals for the location of QTLs.

Having mapped a QTL to a likelihood interval, the location of the QTL might be narrowed down to perhaps 1/50 – 1/100 of a chromosome (about 0.1% of the genome in higher plant like tomato) (Paterson *et al.*, 1990a). With this high resolution gene mapping, it is difficult to prove that a QTL represents only one gene. It is generally acknowledged that a typical higher plant genome includes perhaps 10,000 to 1,00,000 genes. Consequently, 0.1% of the genome may include 10-100 genes which represent a QTL. As a result, several genes lying close together, each with a small effect on a trait, may appear as a single QTL of large effect (Michelmore and Shaw, 1988; Paterson, 1988; Paterson *et al.* 1991b). Techniques with higher resolution power are necessary to reveal how many genes actually represent a single QTL.

Dosage Effect of Gene on QTL

Many plants are diploid, having either zero, one or two copies of a particular gene (allele). The phenotypic effect of an allele often depends upon the number of copies present in an individual, and the nature of the alternate allele when two alleles are present together (Paterson *et al.*, 1991b).

An individual QTL can be described by the effects of gene dosage in anyone of the following classes :

(1) **Additivity :** The phenotype of the hybrid (heterozygote) resulting from a cross between two parents differing at one locus falls exactly half way between the parents (Fig 17.1) . Additivity is expressed as the additive effect (a) of an allele on the phenotype which is thus half the difference between phenotype of the parents.

(2) **Dominance/Recessiveness :** The phenotype of the heterozygote is identical to that of one parent and the allele from that parent is designated dominant. The allele of the alternate parent is designated recessive. Dominance is expressed as the dominance deviation (d) of an allele or the difference between the midparent value and the observed phenotype of the heterozygote.

(3) **Overdominance :** The phenotype of the heterozygote is superior to that of either parent. In this case, the additive effect is zero (since the parents are the same) but the dominance deviation is significant.

(4) **Underdominance :** It is the reciprocal of overdominance. The phenotype of the heterozygote is inferior to that of either parent and the dominance deviation is negative.

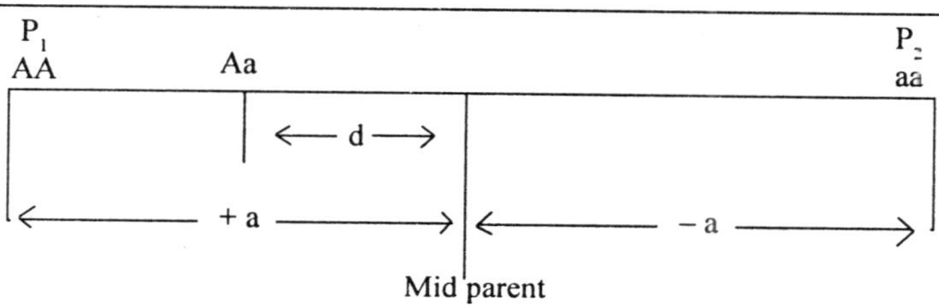

Fig. 17.1: Dosage effect of gene

Procedures have been worked out in detail to study the effects of gene dosage at a QTL using genetic markers (Mather and Jinks, 1971; Edwards *et al.*, 1987). Three fundamental steps are involved. Firstly, genetic markers near an allele of interest are analysed to determine whether an individual carries 0, 1, or 2 copies of the allele. Secondly, the average phenotypes of individuals carrying 0,1 or 2 copies of the allele are determined. Finally, statistical techniques are used to determine whether the change in phenotype per additional copy of the allele most closely fits additivity, dominance/ recessiveness, or overdominance (Paterson *et al.*, 1991b).

The ideal population for studying dosage effect of gene at a single QTL in diploids is an F_2 population segregating in the classical 1 : 2 : 1 Mendelian ratio for the three possible genotypes (AA : Aa : aa).

Individual QTLs have been described which show additive (d = 0), dominant (d = a), recessive (d = – a) and overdominant (a = 0, d > 0) gene action (Paterson *et al.*, 1990a, 1991a). Some QTLs cannot be assigned clearly to any of these classes as they appear intermediate between two classes (*e.g.* partial dominance where a > 0 and d > 0).

Phenotypic Effect and Environmental Sensitivity of QTL

A single QTL sometimes influences two or more phenotypic traits (pleiotropy). Such traits show association with each other which is difficult to separate. On the other hand, the association between different traits may be due to several linked QTLs which can be separated by recombination. A QTL may include 10-100 genes when the location of a QTL is narrowed down to about 0.1% of the genome. Consequently, the QTLs which map at different locations can be easily inferred to represent different genes. But the QTLs which map to similar locations cannot be assumed to represent the same gene and may be also different. With the development of further saturated genetic map, it will be possible to analyse the phenotypic effect of each QTL.

In plant breeding experiments, different genotypes do not perform equally well in different environments. Some genotypes show better performance in some specific environments while others perform better in other environments. Sometimes these differences among genotypes can be attributed to relatively simply inherited traits such as photoperiod response, vernalization requirement, susceptibility to particular strains of pest etc. (Paterson *et al.*, 1991b). But the differences in adaptation of plant genotypes to particular environments may also be due to environment-sensitive QTLs. From a study of 29 QTLs mapped in three different environments, Paterson *et al.*, (1991a) observed that 4 QTLs (14%) expressed in all three environments, 10 QTLs (34%) expressed in two environments and 15 QTLs (52%) expressed only in a single environment.

Environmental sensitivity of QTLs may be useful to the breeder. It will be easier for the breeder to work with the QTLs which express consistently in different environments. Despite this, the breeder may attempt to develop a variety by combining several QTLs with different environmental sensitivity. Such variety may perform better over a wide area because of its buffering against the vagaries of environment.

QTL Analysis Procedure

The procedure of QTL analysis is simple and involves several steps:

(1) The parents which show high levels of polymorphism and large differences for the character to be mapped are selected.

(2) The parents are hybridized to develop desirable generations *e.g.* F_2, F_3, BC (backcross), DHs (double haploids or dihaploids), RILs (recombinant inbred lines).

DHs are completely homozygous plants produced by doubling the chromosome number of pollens of F_1 plants. DHs are genetically different as they carry different parental chromosome combinations. RILs are derived by single-seed descent method from selfing of individual F_2 or backcross plant followed by atleast six generations of selfing. Each RIL has a characteristic combination of genes with a different pattern of alternative alleles at multiple loci.

(3) The individual plants/lines of the desirable generation are characterised for molecular markers like RFLPs, isozymes and microsatellites etc. The plants/lines are evaluated for the character in one or several environments.

(4) The data are analysed and the QTLs for the character are mapped.

In practice, a cross is made between two plants that are genetically different for one or more characters of interest. Segregating progeny are obtained from the hybrid (commonly F_2, backcross or recombinant inbred lines). A number of progeny (usually more than 100) are evaluated for the character of interest and for their genotypes at RFLP marker loci at regular intervals of 10-20 cM (centi Morgan ; 1 cM is equal to 250-300 kb in rice) throughout the genome. A search is then made for associations between the segregating RFLP markers and the character of interest. If such associations are found, they should be due to linkage of the

RFLP marker to a gene(s) affecting the character (Tanksley *et al.*, 1989).

Statistical Approaches for QTL Mapping

Several statistical approaches have been proposed to map QTL and to estimate their effects. Although different approaches involve different degrees of statistical sophistication, they all have similar reliability. Two commonly used methods of QTL mapping are :

(1) **Marker Regression :** This method involves standard linear regression with phenotype as dependent variable and marker as independent variable.

(2) **Interval Mapping by Multiple Regression :** This method uses flanking markers for interval mapping using individual observations rather than marker means. Two different statistical procedures are commonly used :

(a) Maximum likelihood procedure, and
(b) Weighted least squares procedure.

QTL mapping using the above two approaches has been discussed in detail by Kearsey and Pooni (1996).

Factors Affecting QTL Analysis

The ability to detect a QTL with a marker (say, RFLP) depends on

(1) magnitude of the QTL's effect on the character,
(2) size of the segregating population, and
(3) recombination frequency between the marker and the QTL.

When a RFLP marker and a QTL are far apart on a chromosome, occurrence of recombination reduces the probability (likelihood) that the marker and the QTL will be transmitted to the same progeny individuals. On the other hand, small distance between marker and QTL increases the probability of their transmittance to the same individuals. As a consequence, analysis based on single RFLP markers may not detect the QTL with statistical significance or its effect may be underestimated. The approach of interval mapping overcomes this problem by basing analysis on two linked markers enclosing an interval which may contain a QTL (Tanksley *et al.*, 1989).

The number of QTLs affecting a character depends upon-

(1) nature of the quantitative trait,
(2) nature of the parents, and
(3) phenotypic variance of the character.

The small effect of QTLs on the character can be detected by the marker regression approach and it depends on several factors (Brar and Dhaliwal, 1997) :

(1) **Map distance from the nearest marker to QTL :** The closer a QTL is to a marker, the smaller the effects of QTL can be detected.

(2) **Size of segregating population :** The effects of lesser QTL will reach statistical significance with increase in the population size.

(3) **Heritability :** If heritability of the character is low, a QTL may not be detected.

In experiments with smaller population sizes, and higher probability thresholds, QTLs that describe less than 3 per cent of the phenotypic variance are not normally detected (Tanksley, 1993).

The most important consideration in interpretation of QTL data is the threshold employed for inferring a QTL to be statistically significant. Significant thresholds of 99.8% or more for any single QTL and 95% for all QTLs across a genome are usually necessary (Paterson, 1995).

QTLs showing Epistasis

Epistasis is the interaction between alleles of different genes in a genome. It is also called non-allelic interaction or inter-allelic interaction. QTLs show epistasis. Epistasis involves two or more loci unlike one locus involved in intra-allelic interaction showing dominance/recessive relationship.

Until recently, QTL mapping experiments have provided very little evidence in support of the presence of epistasis among QTLs but sometimes its presence at a statistically nonsignificant level. However, hints of epistasis among QTLs have come from the demonstration of "genetic background effects" in quantitative genetics of rice (Kinoshita and Shinbashi, 1982 ; Sato and Sakamoto 1983), tomato (Tanksley and Hewitt, 1988) and *Drosophila.* Evidence of epistasis between QTLs in the genetic control of several agronomic traits came from the study of recombinant inbred lines (RILs) in soyabean (Lark *et. al.*, 1995).

But the absence of epistasis in previous QTL mapping studies might be due to insufficient replication and/or insufficient statistical resolution to detect epistasis in the presence of many QTLs with large main effects (Paterson, 1995). Detection of epistasis through QTL mapping may account for a portion of the "genetic difference between parents". The effects of some QTLs appear independent of each other. But it is becoming clear that the "whole effect" of all QTLs influencing a character is greater than the sum of individual QTL effects thereby indicating the presence of epistasis.

It is worthwhile to mention here that the analysis for presence of epistasis is more subject to the problem of false positive results than the analysis for individual QTL's effect.

Comparative Analysis of QTLs

Comparative mapping is the alignment of the chromosomes of related species on the basis of genetic mapping of common DNA markers. It has revealed that many distantly related species have co-linear maps for portion of their genomes.

Comparative mapping has shown a close correspondence among QTLs affecting complex traits such as seed size, day-neutral flowering in different taxa (species) *viz.* rice, wheat, maize, barley, sugarcane and sorghum (Paterson *et al.*, 1995b). A close correspondence among QTLs affecting height in maize, sorghum and other grasses has been reported (Lin *et al.*, 1995). Besides this, comparative analysis has demonstrated a close correspondence of QTLs on duplicated chromosome segments within a particular species. It indicated that chromosome duplication might contribute to polygenic inheritance (Paterson, 1995).

Comparative mapping helps to :

(1) analyse genomes (many DNA markers can be used among different species of the same family and thereby increase the number and map density of genetic markers for genome analysis).
(2) determine colinearity and synteny among genomes (synteny refers to the presence of two or more genes on the same chromosome without any information of linkage);
(3) identify orthologous QTL; and
(4) explore the possibility of closing orthologous QTL.

Benefits of QTL Correspondence

Close correspondence of QTLs affecting a quantitative character among diverse species offer many benefits.

(1) Close correspondence indicates that QTL analysis in one species may predict results in other species.
(2) Correspondence of QTLs among different species provides support for the use of model systems in research on complex phenotype. For instance, isolation and characterization of QTLs from small roots and inflorescence of a particular species may be difficult. In such cases, corresponding QTLs may be isolated easily from the enlarged root of turnip, inflorescence of

cauliflower or fruit of tomato to study specific aspects of plant growth and development.

(3) Correspondence among QTLs in different species supports the hypothesis that a relatively small number of mutations (genes) accounts for a large portion of phenotypic variation in many populations. Such correspondence seems unlikely if the number of mutations (genes) were large and equally probable.

Cloning QTLs and Chromosome Walking

Cloning genes on the basis of chromosomal location or map position (called map-based gene closing) is easy for those genes which produce discrete phenotypes. In other words, oligogenes governing qualitative characters are very convenient for map-based cloning. But QTLs influencing traits of economic importance or agricultural productivity are difficult to clone by traditional techniques because for most QTLs, only the phenotype is known and not the QTL product itself. Despite this, QTLs produce a relatively small effect on phenotype and these small effects can be concealed by the effects of other genes or by nongenetic factors. Cloning QTLs is difficult because their biochemical functions are unknown.

> **Gene cloning** - It is the production of exact replicas of a single gene.
> **Clone** - It is a group of genetically identical cells or organisms produced from a single cell or organism by mitosis.

Using even the latest "second-generation" experimental designs based on progeny testing, a resolution of about 3 cM has been achieved in QTL mapping. In the genome of several crop plants, 3 cM represents nearly 0.1% of the genome containing perhaps 100 or more genes. 1 cM represents about 600 kb of DNA (in tomato). Therefore, 3 cM represents nearly 1800 kb of DNA.

Some recent developments in molecular biology have made cloning of QTLs easier. The technology for cloning large segment of intact chromosomal DNA (> 10^5 bp of DNA) has been developed.

It has facilitated "chromosome walks" (Steinmetz *et al.*, 1982). A plasmid or cosmid vector can incorporate a foreign DNA segment of 15-30 kb. But artificial chromosomes (Burke *et al.*, 1987) carrying large foreign DNA fragment, nearly 10 times the foreign DNA of plasmid or cosmid vector, have been developed in yeast and bacteria. These are called yeast artificial chromosome (YAC) and bacterial artificial chromosomes (BAC) respectively.

By randomly making a large number of yeast artificial chromosomes (YACs) from a genome of interest, it is possible to identify a YAC near the QTL of interest by using the closest RFLP as a probe. In the next step, the first YAC is used as a probe to identify a second YAC which overlaps with and also extends beyond the first YAC. This is referred to as "chromosome walking" . Repeating this several times, a series of YACs which lie end to end can be identified which actually stretch over the QTLs region of interest. Subsequently, DNA vectors each of which carries a smaller bit of the QTLs region of interest (cosmid, phage or plasmid) are introduced into separate plants lacking the trait of interest. Those bits which cause the plant to exhibit the trait of interest are inferred to harbour the target QTL. A strategy for cloning a gene, which is known by its effect on phenotype, using "chromosome walking", has been shown (Fig. 17.2).

Direct isolation of genes from megabase DNA clones is becoming more efficient (Lovett et al., 1991; Collins, 1992 and Morgan et al., 1992). Several PCR based techniques are now available for identifying genes expressed specifically in particular tissues, developmental stages or genotypes. Identification of discrete mutations corresponding to QTLs offer a more rapid means to dissect quantitative traits than do breeding approaches (Paterson et al., 1990a; Dorweiler et al., 1993).

QTLs Mapped in Rice

Several QTLs influencing traits of economic importance have been mapped in rice genome. Xiao *et al.*, (1996) analysed BC_2 test cross families from the interspecific cross (*O. sativa* X *O. rufipogon*)

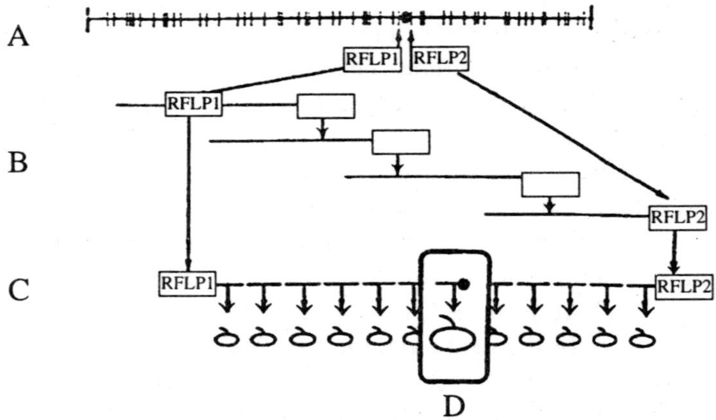

Fig. 17.2: A strategy for cloning a gene, known only by its effect on phenotype, using "chromosome walking". Starting with the knowledge that a gene lies between two mapped DNA clones (A), one restriction fragment length polymorphism (RFLP) clone can be used to identify a YAC which includes the clone, by screening a large number of candidate YACs (usually a library of 10^5 or more.) A technique called inverse PCR (Ochman *et. al.* , 1988) can be employed to obtain a small bit of DNA lying at the end of a YAC. This terminal DNA is then used as a probe to identify a second YAC, and the cycle is repeated until a YAC is identified which includes the second RFLP. In practice, one seldom has the luxury of knowing which direction to walk in, and occasionally the required YAC is not present in the library for any number of reasons; thus, this walking experiment is invariably more difficult than it sounds. In principle, however all the genomic DNA which lies between the two starting DNA clones might be identified. In another laborious experiment, all of this genomic DNA can then be introduced, a small bit at a time (C), into plants lacking the trait of interest (Schell, 1987). Those bits which cause the plant to exhibit the trait of interest (D) are inferred to harbour the target gene, and may be dissected further (After Paterson *et al.*, 1991b).

and found that *O. rufipogon* alleles at marker loci RM5 on chromosome 1 and RG 256 on chromosome 2 were associated with increased yield. The phenotypic effect of the lines carrying *O. rufipogon* alleles at these two loci corresponds to an 18 per cent and 17 per cent yield increase over *O. sativa* parent. These results suggest that molecular markers can be used to identify QTL from wild species responsible for transgressive segregation. The experimental population for this QTL analysis was developed as given in Fig. 17.3

Oryza sativa x *O. rufipogon*
(V_{20A})
↓
F_1 X
O. sativa
↓ (V_{20B})
52 BC_1 plants produced; of these, only
10 Best BC_1 plants selected X *O. sativa*
(V_{20B})
↙
3000 BC_2 plants produced,
300 Best BC_2 plants selected X Variety (Ce64)
↙
300 BC_2 Test cross families
i) Evaluated for yield and yield components
ii) Assayed for RFLPs and QTLs mapped

Fig. 17.3: Procedure for developing experimental populations for QTL analysis in rice (Xiao *et al.*, 1996)

In another experiment, the populations for QTL analysis in rice were developed as given in Fig. 17.4 by Xiao *et al.* Some examples of QTLs mapped using molecular markers in rice have been presented in Table 17.1.

Future Priorities in QTLs Research

QTLs research, as an adjunct of quantitative genetical analysis, should continue in the coming years in several avenues :

(1) Fine mapping of QTLs should be done so as to determine whether they are single genes or a cluster of tightly linked genes .
(2) The QTLs which perform consistently over different environments should be identified for increasing

agricultural productivity. Individual QTLs show a range of environmental sensitivities.

(3) Complementary QTLs which contribute towards transgressive segregation must be identified from the interspecific crosses.

(4) Efforts must be concentrated for map-based cloning of QTLs to improve characters of economic importance in crop plants.

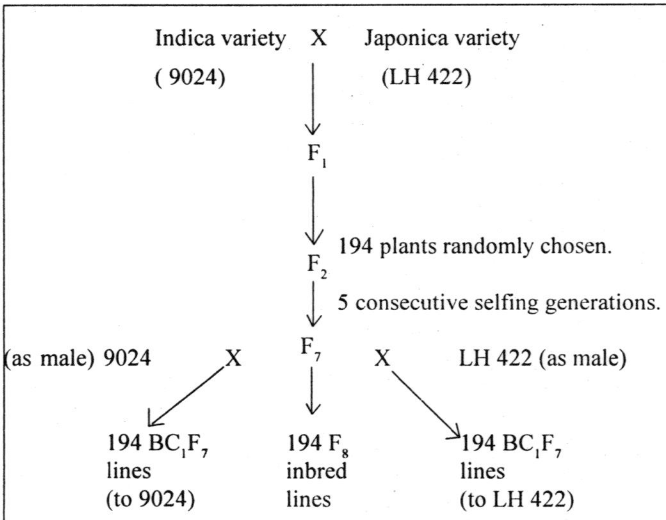

Fig. 17.4: Diagram showing procedures for developing experimental populations for QTL analysis

Table 17.1: Some examples of QTLs mapped using molecular markers in rice

Trait	Population used	No. of QTLs mapped	QTLs	Reference
Plant height	BC	5	ph 2, ph 3, ph 5, ph 6, ph 8	Xiao *e. al.*, 1996
Days to heading	BC	5	dth 3-1, dth 3-2, dth 4, dth 7, dth 8	-do-
Days to maturity	BC	3	dtm 4, dtm 7, dtm 8	-do-
Panicle length	BC	2	pl 5, pl 9	-do-
Panicles per plant	BC	1	ppp 4	-do-
Spikelets per panicle	BC	2	spp 3, spp 5	-do-
Grains per panicle	BC	3	gpp 3, gpp 4, gpp 5	-do-
Percent seed set	BC	2	pssr 6, pssr 7	-do-
1000-grain weight	BC	5	gw 3, gw 4, gw 5, gw 7, gw8	-do-
Spikelets per plant	BC	3	sppl 3, sppl 5, sppl 11	-do-
Grains per plant	BC	4	gppl 3, gppl 4, gpp 5, gppl 11	-do-
Grain yield	BC	2	gy 8, gy 11	-do-
Seedling vigour	F_2	4	—	Redona and Mc Kill, 1996
Blast resistance	F_4	4	br 2, br 4, br 9, br 12	Fukuoka and Okuno, 1997a
Submergence tolerance	RIL	4	—	Brar and Dhaliwal, 1997
Root/Shoot ratio	RIL	9	—	-do-
Hybrid breakdown	F_2	2	hwd 7, hwd 10	Fukuoka and Okuno, 1997b

Individual QTL is named by abbreviation of the concerned trait and chromosome number (1-12). In cases where multiple QTLs affecting a trait are found along the same chromosome, the QTLs are distinguished by numbers or letters indicating the order in which the QTLs were discovered (dth 3-1, dth 3-2).

REFERENCES

Brar, D.S. and Dhaliwal, H.S. 1997. Molecular markers and their application in crop improvement. Third Agricultural Science Congress, 12-15 March, Ludhiana.

Burke, D.T., Carle, G.F. and Olson , M.V. 1987. Cloning of large segments of exogenous DNA into yeast by means of artificial chromosome vectors. *Science,* 236: 806-812.

Collins, F.S. 1992. Positional cloning : let's not call it reverse any more. *Nature Genet.* 1: 3-6.

Dorweiler, J., Stec, A., Kermicle, J., and Doebly, J. 1993. Teosinte glume architecture - 1: A genetic locus controlling a key step in maize evolution. *Science,* 262 : 233.

East, E.M. 1915. Studies on size inheritance in *Nicotiana. Genetics,* 1: 164-176.

Edwards, M.D., Stuber, C.W. and Wendel, J.F. 1987. Molecular marker facilitated investigations of quantitative trait loci in maize .I. Numbers, genomic distribution and types of gene action. *Genetics,* 116 : 113 - 125.

Fukuoka, S. and Okuno, K. 1997a. QTL analysis for field resistance to rice blast using RFLP markers. *Rice Genet. Newsl.,* 14:98-99.

Fukuoka, S. and Okuno, K. 1997b. RFLP mapping of complementary genes for hybrid breakdown in cultivated rice. *Rice Genet Newsl.,* 14 : 99-102.

Geldermann, H. 1975. Investigations on inheritance of quantitative characters in animals by gene markers. I. Methods. *Theor. Appl. Genet.,* 46 ; 319-330.

Johanssen, W. 1909. *Elemente der exakten Erblichkeitsllehre.*

Kearsey, M.J. and Pooni, H.S. 1996. The Genetical Analysis of Quantitative Traits. Chapman and Hall, London, 133-164.

Kinoshita, T. and Shinbashi, N. 1982. Identification of dwarf genes and their character expression in the isogenic background. *Japanese J. Breeding,* 32 : 219-231.

Lander. E.S. and Botstein, D. 1989. Mapping Mendelian factors underlying quantitative traits using RFLP linkage maps. *Genetics,* 121 : 185-199. Corrigendum, Genetics, 136 : 705.

Lark, K.G., Chase, K., Adler, F., Mansur, L.M. and Orf, J.H. 1995. Interactions between quantitative trait loci in soyabean in which trait variation at one locus is conditional upon a specific allele at another. Proceedings Natl. Acad. Sci., 92 : 4656-4660.

Lin, Y.R., Schertz, K.F. and Paterson, A.H. 1995. Comparative analysis of QTLs affecting plant height and maturity across the poaceae, in reference to an interspecific sorghum population. *Genetics,* 141 : 391 - 411.

Lovett, M., Kere, J. and Hinton, L.M. 1991. Direct selection: A method for the isolation of cDNAs encoded by large genomic regions. Proceedings Natl. Acad. Sci., 88 : 9628 - 9632.

Mather, K. and Jinks, J. L. 1971. *Biometrical Genetics.* Cornell University Press, Ithaca, New York.

Michelmore, R.W. and Shaw, D.V. 1988. Character dissection. *Nature,* 335 : 672-673.

Morgan, J.G., Dolganov, G.M., Robbins, S.E., Hinton, L.M. and Lovett, M. 1992. Nucleic Acids Res., 20 : 5173-5179.

Nilsson-Ehle, H. 1909. *Kreuzunguntersuchungen an Hafer und Weizen.* Lund.

Ochman, H., Gerber, A.S. and Hartl, D.L. 1988 Genetic applications of an inverse polymerase chain reaction. *Genetics,* 120 : 621-623.

Paterson, A.H., Lander, E.S., Hewitt, J.D., Paterson, S., Lincoln, S.E. and Tanksley, S.D. 1988. Resolution of quantitative traits into Mendelian factors by using a complete map of restriction fragment length polymorphisms. *Nature,* 335 ; 721 -726.

Paterson, A.H., Deverna, J.W., Lanini, B. and Tanksley, S.D. 1990a. Fine mapping of quantitative trait loci using selected overlapping recombinant chromosomes in an interspecies cross of tomato. *Genet.,* 124 : 735 - 742.

Paterson, A.H., Damon, S., Hewitt, J.D., Zamir, D., Rabinowitch, H.D., Lincoln, S.E., Lander, E.S. and Tanksley, S.D. 1991a. Mendelian factors underlying quantitative traits in tomato : comparison across species, generations and environments. *Genetics,* 127: 181-197.

Paterson, A.H., Tanksley, S.D. and Sorrells, M.E. 1991b. DNA markers in crop improvement. Advances in Agronomy, ed. by D.L. Sparks, Academic Press, New York, 39-90.

Paterson, A.H. 1995. Molecular dissection of quantitative traits: progress and prospects. *Genome Research,* 5 : 321-333.

Paterson, A.H., Lin, Y.R., Lio, Z., Schertz, K.F. , Doebly , J. F. , Pinson, S.R.M., Liu, S.C., Stansel, J.W., and Irvine, J.E. 1995b. Convergent domestication of cereal crops by independent mutations at corresponding genetic loci. *Science,* 269 : 1714 - 1718.

Redona, E.D. and Mackill, D.J. 1996. Mapping quantitative trait loci for seedling vigour using RFLPs. *Theor. Appl. Genet.,* 32 : 395-402.

Sato, S. and Sakamoto, I. 1983. Inheritance of heading time in isogenic line rice cultivar, Taichung 65 carrying earliness genes from a reciprocal translocation homozygote, T3 - 7. *Japanese J. Breeding,* 33 : 118-119.

Sax, K. 1923. The association of size differences with seed coat pattern and pigmentation in *Phaseolus vulgaris. Genetics,* 8 : 552-560.

Schell, J. 1987. Transgenic plants as tools to study the molecular organization of plant genes. Science, 237 : 1176 - 1182.

Steinmetz, M., Minard, K., Horvath, S., Mc.Nicholas, J. Srelinger, J., Wake, C., Long, E., Mach, B. and Hood, L. 1982. A molecular map of the immune response region from the major histocompatibility complex of the mouse. *Nature,* 300 : 35 - 42.

Tanksley , S.D. and Hewitt, J.D. 1988. Use of molecular markers in breeding for soluble solids in tomato : a re-examination. *Theor. Appl. Genet.* 75:811-823.

Tanksley, S.D., Young, N.D., Paterson, A.H. and Bonierbale M.W. 1989. RFLP mapping in plant breeding : new tools for an old science. *Biotechnology,* 7 : 257 - 264.

Tanksley, S.D. 1993. Mapping polygenes. *Annual Review of Genetics,* 27 : 205 - 233.

Thoday, J.M. 1961. Location of polygenes. *Nature,* 191 : 368-370.

Xiao, J., Ahn, S.N.G. , McCouch, S.R., Tanksley, S.D., Li, J. and Yuan, L. 1996. Genes from wild rice improve yield. *Nature,* 384 : 223-224.

18

Rice Biotechnology and Genetic Engineering

Introduction

In order to feed the ever increasing human population, new rice varieties are needed with increased yield potential, superior grain quality, enhanced resistance to diseases and pests, and greater tolerance to natural abiotic stresses such as cold, drought and nutrient deficiencies. Biotechnology appears as perhaps the most important new resource in meeting this challenge of developing new rice varieties. The techniques of rice biotechnology are derived from plant tissue culture and molecular biology. Genetic engineering is basically a combination of tissue culture and molecular techniques.

Rice biotechnology and genetic engineering have great potential in accelerating the pace of conventional breeding programmes. Over the last 20 years, rice biotechnology has made significant strides in several avenues. These include :

(1) Tissue culture
   a. Embryo rescue
   b. Anther culture

(2) Molecular marker—aided selection
(3) RFLP map construction of rice genome
(4) Transformation system

a. Protoplast transformation
b. Biolistic transformation

(5) Gene tagging
(6) DNA finger printing of pathogens and pests
(7) Germplasm management
(8) Transgenic plant production

Embryo Rescue

Many wild species of the genus *Oryza* fail to hybridise with cultivated species because of the incompatibility. This incompatibility arises from differences in the number or the genetic constitution between chromosomes of the two species. Sometimes, fertilization takes place but the embryo fails to develop. In such cases, embryo rescue technique has been found to be very useful. This technique involves isolation and culture of immature embryos on nutrient media in a test tube and subsequently regeneration of the embryo into a whole plant called distant hybrid or wide hybrid.

Initially the hybrid embryo contains chromosomes higher than diploid (2n = 24) but it is possible to maintain the hybrid embryo through tissue culture until the chromosome number declines to that of cultivated rice (2n = 24) and fertility is restored. The 12 chromosomes of the haploid set of such lines, derived from the hybrid embryo, closely resemble the chromosomes of the cultivated parent. But small regions of their chromosomes are replaced by segments of the wild genome through random cross-over events (Bennett, 1994).

Useful genes for disease and insect resistance have been transferred using this technique from *Oryza officinalis* and *O. australiensis* into rice (Jena and Khush, 1990). However, genomes of some wild species of rice are strongly non-homologous and gene transfer through conventional breeding method such as backcrossing is highly restricted. Therefore, embryo rescue technique could be employed to explore the possibility of getting a distant hybrid with desirable characteristics.

Anther Culture

New rice varieties may be produced by repeated selfing of a hybrid. At least 5 selfing generations are necessary to generate a stable rice line by this method. But in regions where only one rice crop is grown per year, it will at least require 6 years to develop the homozygous plants. But this time period can be reduced by using anther culture technique which produces the homozygous plants in two generations.

In this method, anthers are removed from the floret of the hybrid, sterilised and placed on a suitable nutrient medium. Haploid callus (irregular mass of undifferentiated cells) develops from the pollens located inside the anthers. The haploid cells either spontaneously diplodise or are treated with colchicine for diplodisation to form a dihaploid plant. The dihaploids are completely homozygous (carrying same alleles in two chromosomes at a locus) at all loci.

Anther culture provides the advantage of early generation selection for a trait governed by recessive genes. Cold tolerance is one of the traits introduced into cultivated indica varieties by use of this method at IRRI.

Molecular Marker-Aided Selection

Numerous genes of economic importance such as disease and pest resistance are repeatedly transferred from one varietal background to another by backcross method of conventional breeding. Most genes behave in a dominant-recessive manner and require time consuming efforts to transfer (Khush, 1996). Screening procedures for these genes are sometimes cumbersome and expensive and require large space.

Transferring a recessive gene by repeated backcrossing is more difficult than transferring a dominant gene across varieties. In the transfer of a recessive gene, one generation of selfing is required after each backcross so that the recessive gene in homozygous condition can manifest phenotypically permitting selection. If only one generation is grown per year in the absence of off-season nurseries and glass houses, the procedure requires at least 9 years to transfer even a dominant gene, keep aside the recessive gene transfer. The procedure becomes highly complicated when the breeder is interested in screening several recessive genes at a time governing different characters.

In such situations, breeders like to practise marker aided selection. If the desirable genes can be tagged with a marker (either traditional or molecular marker), the presence or absence of the associate molecular marker would indicate the presence or absence of the gene at a very early stage. Screening of the marker is easier than that of the gene. For example, in some traditional rice varieties of north-east India, a gene for resistance to the important insect pest brown plant hopper (Bph) resides close to a gene specifying a traditional marker i.e. purple coleoptile colour. When a resistant plant with a purple coleoptile is hybridized to a susceptible plant with a green coleoptile, more than 95% of F_2 plants showing purple coleoptile colour are also resistant to Bph (Bennett, 1994). In this case, coleoptile colour is a traditional morphological marker for selecting the Bph resistance gene.

Unfortunately, few traditional/morphological markers are known and they tend to be specific for particular rice varieties. Moreover, the marker and the gene must be either both dominant or both recessive. Using morphological markers, only single major genes can be screened and selected but it is not applicable to many of the most interesting and agronomically interesting traits which are quantitative in nature and governed by numerous unlinked genes.

The discovery of molecular markers like isozyme and DNA markers has added enormously to the efficiency of marker-aided selection breeding. DNA and isozyme markers are more numerous than morphological markers and are available for any cross. Molecular markers are codominant and permit observation of both parental markers in the hybrid as there is no dominance-recessive behaviour. Moreover, molecular markers are not affected by the environment and do not interact with other genes. But scoring procedure of molecular markers is more difficult and expensive than that of morphological markers. The most commonly used DNA markers are restriction fragment length polymorphisms.

In rice, more than 40 genes encompassing disease and insect resistance, fertility restoration, photoperiod and thermoperiod sensitive male sterility, wide compatibility, dwarfism and submergence tolerance have been tagged with molecular markers. A molecular map of rice having more than 1000 DNA markers has been prepared largely by the efforts of the research group of Cornell University (Tanksley *et al.*, 1992) and Rice Genome mapping project of Japan (Sasaki, 1994).

At IRRI, selection approaches using molecular markers are being tried in the breeding programme of irrigated rice varieties. High cost of the chemicals required in molecular marker technology greatly reduces the wide-spread use of this technology in rice improvement programmes of most developing countries.

RFLP Map Construction of Rice Genome

More than 1000 RFLP markers have been placed on the genetic map of rice through co-segregation analysis. Traditional morphological markers and isozyme markers are also being placed on the map (Bennett, 1994).

RFLP map construction in rice has become a reality due to several factors namely:

(1) rice has a small haploid genome of 0.6 picogram,
(2) it is a true diploid,
(3) it has a relatively little repetitive DNA sequences and they do not interfere much with map construction,
(4) research on rice biotechnology has a global network of collaborating laboratories which freely exchange scientific knowledge and breeding materials.

The first mapping population employed at Cornell University in the United States was an F_2 population derived from an indica/tropical japonica intraspecific cross. Tropical japonica varieties are also known as javanica varieties. The map of Cornell University has been refined with a second mapping population consisting of vegetatively propagated inbred plants derived from an *Oryza sativa /Oryza longistaminata* interspecific cross. Other mapping populations developed in China and Japan are being reconciled with Cornell map (Bennett, 1994).

Protoplast Transformation System

Protoplasts are naked protoplasms surrounded by plasma membranes. Protoplasts are obtained by digestion of plant cell wall with enzymes like cellulase, hemicellulase and pectinases. Protoplasts are easily amenable to transformation as their plasma membranes allow the entry of foreign DNA.

Using molecular techniques, protoplasts of two non-hybridizable species can be cultured together and fused to form a somatic hybrid of two species. Such fusions require a variety of treatments like polyethylene glycol (PEG), sodium nitrate, high P^H and calcium, inactivated viruses, polyvinyl alcohol, antibodies, plant lectins, gelatin and electric stimulation (Chopra and Narasimhulu, 1990). Following fusion, the protoplast population consists of a heterogeneous mixture of homokaryons and heterokaryons. The heterokaryons constitute a small proportion of the mixture. They are selected either visually or using biochemical selectable markers and, after cell wall synthesis, regenerated as whole plants called somatic hybrids.

Protoplast culture techniques have enabled the introduction of alien genes from other species into rice. It also permits the re-introduction of rice genes that have been extracted and modified to give altered phenotypes of commercial importance.

In rice cells, there are no natural vectors for DNA transfer (Uchimiya and Toki, 1994). Moreover, *Agrobacterium* mediated gene transfer which is commonly used in dicots cannot be applied to monocots like rice because the bacterium cannot form galls. So transformation in rice has been achieved mainly through protoplasts using direct DNA transfer methods.

For successful gene transfer in rice, DNA uptake through the protoplast membrane is mediated either by addition of the neutral polymer PEG or by brief exposure to an electric current (electroporation) in several laboratories including IRRI.

The key to success with protoplast transformation in rice lies with regeneration of viable and fertile plants. The first report on plant regeneration from protoplasts of japonica rice was published in 1985. Protoplasts prepared from suspension cells appear to give best regeneration frequencies. Japonica varieties respond better than indica varieties in regeneration ability. Chromosomal abnormalities may occur during the 4-6 month period of tissue culture required to produce suitable suspension cells from explants and then to obtain regeneration of protoplasts into callus and whole plants. Regeneration ability also decreases with the increase in time period of tissue culture.

Biolistic Transformation System

Many alternative techniques of rice transformation have been reported but they have not yet proved as successful as the protoplast-based methods. One of these techniques is the biolistic approach. In this technique, microscopic gold particles coated with foreign DNA are fired at callus or intact plant explants such as embryos. The particles are accelerated by gunpowder, compressed helium gas, compressed air or electric discharge. These fast moving particles penetrate several cell layers and then rest either in the space between two cells or inside plant cells. Being inside the plant cells, the foreign DNA may be active in transcription and may give rise to transient expression of the foreign DNA. Sometimes, the foreign DNA integrates itself into host chromosome and stably transforms the host cell.

The transformed host cell multiplies mitotically to produce a clone of transformed cells which regenerate to form a transformed plant. If the transformed rice plant produces flowers, the foreign DNA will be inherited from one generation to the next through sexual reproduction. If the transformed plant undergoes reproductive phase, it can be maintained either asexually for research needs or by tissue culture techniques. These transformed plants are also called transgenic plants.

The biolistic method is a very recent technique and it requires refinements before its extensive use in rice transformation. Biolistic transformation of both japonica and indica varieties was achieved at IRRI in 1993.

Gene Tagging

RFLP map of rice genome can be used to locate and tag genes of economic importance such as disease and pest resistance, wide compatibility etc. More than 40 rice genes have already been tagged with DNA markers.

If different major genes specifying resistance to several insect pests of rice are tagged with RFLP markers, it would be possible to follow the inheritance of all these genes simultaneously without any cumbersome entomological test under field condition for scoring resistant plants. This would be particularly valuable in a breeding programme of developing rice varieties having resistance to several insects.

Tagging genes without reference to a map is also possible by using a technique called "bulked segregant analysis". This is particularly powerful for tagging a major gene (Bennett, 1994). It may be used with RFLP markers or RAPD (Randomly Amplified Polymorphic DNA) markers. The first RFLP or RAPD marker which shows co-segregation with the particular gene of interest may not be the closest marker.

DNA Finger Printing of Pathogens and Pests

Many pathogens and pests of rice show considerable genetic diversity in terms of virulence and host range. Breeders are always interested in screening germplasm against virulent pathogens and insect pests. RFLP and RAPD analysis of the genomes of rice pathogens and insect pests have added to our understanding of how pathogens and pests evolve and how new resistant rice varieties may be developed using these information.

Germplasm Management

Molecular markers are used to obtain information regarding genetic diversity, to classify and to establish phylogenetic relationship in rice germplasms. Well characterised germplasms are the invaluable assets to the breeders for using them more effectively as donors in result oriented, time-bound breeding programmes for evolving new varieties.

Transgenic Plant Production

When one or more foreign genes are stably integrated into host plant DNA for manifestation as protein, the plant is referred to as transgenic.

Although the production of transgenic rice plants has been reported by several workers (Battraw and Hall, 1990; Datta *et al.*, 1990; Tada *et al.*, 1990; Zhang *et al.*, 1988), the regeneration of fertile and stable transgenic rice plant is yet to be made a routine work. In many plant transformation studies, drug resistance genes such as neomycin phosphotransferase (NPT) or hygromycin phosphotransferase (HPH) have been used as markers. But in a newly developed alternative strategy, herbicide resistance genes such as bialaphos and phosphinothricin (PPT) resistance genes are used as selectable markers.

Bialaphos and PPT are potent herbicides with a wide range of toxicity to numerous crops and weeds. Bialaphos was originally registered as a tripeptide antibiotic produced by the bacterium *Streptomyces hygroscopicus*. Bialaphos consists of PPT, an analogue of L-glutamic acid and two L-alanine residues. Due to removal of the alanine residues by peptidases of plant cells, the PPT inhibits glutamine synthetase enzyme. As a result, ammonia accumulates in plant cell leading to its death. In this way, the herbicidal property is manifested.

The *bar* gene cloned from *Streptomyces hygroscopicus* produces the enzyme phosphinothricin acetyltransferase (PAT) which acetylates the amino terminal group of PPT and thereby abolishes herbicidal activity of PPT. In the case of rice plant, the potential use of *bar* gene as a selectable marker for transformation was indicated by Dekeyser *et al.* (1989). Transgenic rice plant using a CaMV 355 promoter-bar chimeric gene was regenerated. CaMV stands for cauliflower mosaic virus.

Development of transgenic plant with the *bar* gene will make it resistant to the herbicides phosphinothricin and bialaphos. Such transgenic rice plants are expected to give higher yield in the absence of weeds due to herbicide application in rice field, particularly under direct seeding. But these varieties are unlikely to be released in near future because the *bar* gene present in the genome may escape into the weeds through cross-pollination. So the *bar* gene has to be integrated into the chloroplast DNA of rice through biolistic approach. In that case, *bar* gene will not be transmitted through pollen grain which is produced by meiosis of nucleus.

Transgenic rice possessing "bt" endotoxin gene of the bacterium *Bacillus thuringiensis* has been produced in Japan. The plant shows enhanced resistance to lepidopteran pests of rice such as leaf folder. The gene used in this case was not the natural gene from *B. thuringiensis* but a chemically synthesised gene encoding the same amino acid sequence of the natural gene. The synthetic gene had different codons from the natural gene to take care of the differences in the protein synthesising machinery of the rice cell and the bacterium.

In tobacco, transgenic plants with resistance to tobacco mosaic virus (TMV) was developed by incorporating TMV coat-protein gene into tobacco DNA. Although the mechanism of this coat protein cross-protection is poorly understood, yet it has been reported for other positive-strand RNA viruses like rice streak virus. Future research may be undertaken to produce transgenic rice with resistance to two major rice viruses namely Asian tungro virus and American hoja blanca virus.

Gene Silencing in Transgenic Rice

Although *Agrobacterium*-mediated transformation of cereals is now feasible, direct DNA delivery procedures remain in widespread

Table 18.1: Foreign Gene Transfer in Rice Plants

Genotypes	Sources of Protoplast/ materials	DNA transfer methods	Genes transferred
Yamahoushi (J)	CS	Electroporation	NPT II
Taipei 309 (J), Pi4 (J)	CS	PEG	GUS
Taipei 309 (J)	CS	Electroporation	NPT II
Fujisaka 5 (J)	Florets Pathway	Pollen tube	NPT II
Nipponbare (J)	CS	Electroporation	HPH
Yamahoushi (J)	CS	Electroporation	HPH, GUS
Taipei 309 (J)	CS	PEG	HPH
Nipponbare (J)	CS	PEG	HPH
IR 54 (I)	CS	PEG	NPT II
Chinsurah Boro II (I)	CS	PEG	HPH
Taipei 309 (J)	CS	Electroporation	NPT II, GUS
Nipponbare (J)	CS	Electroporation	HPH,GUS

J—Japonica, I—indica, CS—cell suspension,. NPT II—neomycin phosphotransferase II, GUS—ß-glucuronidase.
Courtesy : Uchimiya and Toki (1994)

use for the transformation of monocots. In particular, biolistics, a transformation procedure that introduces naked DNA into plant cells has become the method of choice for cereal transformation in the recent years (Christou, 1996). Although a large number of transgenic plants can be produced in a short span of time, plants regenerated using this method frequently contain multiple, often rearranged copies of the introduced genes.

These multicopy transformants frequently display aberrant segregation of the foreign genes or their expression. Silencing of the *bar* gene (that confers resistance to the herbicide, bialaphos) in transgenic rice is methylation based. Inactivation of the promoter, maize ubiquitin (Ubil), probably leads to silencing of *bar* gene (Kumpatla *et al*., 1997).

Gene silencing can result from the introduction of multiple homologous sequences, a feature common to direct DNA delivery-mediated transformation procedure. These findings have emphasized the need for the careful molecular analysis and multi-generation testing of the transgenic lines before they are subjected to large scale multiplication for commercial cultivation.

Genetic Engineering for Resistance to Sheath Blight

Most plants synthesize an assortment of new proteins as a defense mechanism in response to pathogen invasion or abiotic stress. These include a group commonly referred to as pathogenesis-related (PR) proteins. The PR-proteins of tobacco have been extensively characterised and sub-divided into PR-1, PR-2 (ß-1l, 3 - glucanases), PR-3 (Chitinases), PR- 4, PR-5 (thaumatin like proteins) and PR-6 (protease inhibitors) proteins.

Among the PR-proteins, chitinases which have been found in many species of higher plants have been studied most intensely. They exhibit complex developmental and hormonal regulation (Boller *et al.*, 1983 and Neale *et al.*, 1990). Chitinases hydro-lyse the ß-1, 4 linkages of the N-acetyl-D-glucosamine polymer, chitin, which is a major component of many fungal cell walls. It has been suggested that chitinases may have a protective role against fungal pathogens (Lin *et al.*, 1995). This notion is supported by the absence of chitin in higher plants and by the observation that chitinase has anti-fungal activity *in vitro* .

> **Southern blot analysis** : It is a technique, developed by E.M. Southern, for transferring DNA segments from an agarose gel to a nitrocellulose filter paper via capillary action. Subsequently, the DNA segment of interest is probed with a radioactive, complementary nucleic acid and its position is determined by autoradiography.
> **Western blot analysis** : In this analysis, a specific protein out of a group of proteins separated electrophoretically can be identified. In this case, the probe is a radioactively labelled antibody raised against the protein in question.

In addition, chitinase activity markedly increases in plants after infection with fungal, bacterial and viral pathogens. It can also be induced by UV light and salicylic acid. The constitutive expression of chitinase in cereal plants may have the potential for improving resistance against fungal attack in two ways. Besides its ability to attack fungal cell walls directly, chitinase releases oligo-N-acetyl glucosamines which function as elicitors for activation of defense-related response in rice cells.

Transgenic rice plants which expressed a rice chitinase gene constitutively were developed. Those rice plants with high levels of chitinase exhibited increased resistance to infection by the sheath blight pathogen, *R. solani* .

In that experiment, a 1.1 kb rice genomic DNA fragment, containing a chitinase gene under the control of the CaMV 35S promoter, was cloned into the rice transformation vector pGL2. After transformation of indica rice protoplasts in the presence of polyethyleneglycol, plants were regenerated. The presence of the chimeric chitinase gene in T0 and T1 transgenic rice plants was detected by Southern blot analysis. Western blot analysis of transgenic plants and their progeny revealed the presence of two proteins with apparent molecular weights of 30 and 35 kDa that reacted with the chitinase antibody. Progeny from the chitinase positive plants were tested for their resistance to the sheath blight pathogen, *Rhizoctonia solani* . The degree of resistance displayed by the transgenic plants to this pathogen correlated with the level of chitinase expression.

REFERENCES

Battraw, M.J. and Hall, T.C. 1990. Histochemical analysis of CaMV 35S promoter-ß-glucaronidase gene expression in transgenic rice plants. *Plant Mol. Biol.,* 15:527-538.

Bennett, J. 1994. Currrent status of rice biotechnology in research and application. Plant biotechnology in technical co-operation programmes: documentation report , ed. by G. Fischbeck *et al.*, 90-95.

Boller, T., Gehri, A., Mauch, F. and Vogeli, U. 1983. Chitinase in bean leaves: induction by ethylene, purification, properties and possible functions. *Planta,* 157 : 22-31.

Chopra, V. L. and Narasimhulu , S. B. 1990 . Biotechnology for crop improvement. Genetic engineering and Biotechnology-Concepts, Methods and Applications, Oxford & IBH Publishing Co. Pvt. Ltd, New Delhi, 159 - 180.

Christou, P. 1996. Transformation technology. *Trends Plant Sci.,* 1: 423 - 431.

Datta, S.K., Peterhans, A., Datta, K. and Potrykus , I. 1990 . Genetically engineered fertile indica rice recovered from protoplasts. *Biotechnology,* 8:736-

740.

Dekeyser, R., Claes, B., Marichal, M., Van Montagu, M. and Caplan, A. 1989. Evaluation of selectable markers for rice transformation. *Plant Physiol.*, 90 : 217 - 223.

Jena, K. K. and Khush , G.S. 1990. Introgression of genes from *Oryza officinalis* Well ex Watt to cultivated rice, *O. sativa* L. *Theor. Appl. Genet.*, 80 : 737-745.

Khush, G.S. 1996. Prospects of and approaches to increasing the genetic yield potential of rice. Rice research in Asia : progress and priorities. CAB International and IRRI, Philippines.

Kumpatla, S.P., Teng, W., Buchholz, W.G. and Hall, T. C. 1997 . Gene silencing and reactivation in transgenic rice. *Rice Genet. Newsl.*, 14 : 155-159.

Lin, Willie, Anuratha, C.S., Datta, K., Potrykus, I., Muthukrishnan , S. and Datta, S.K., 1995. Genetic engineering of rice for resistance to sheath blight. *Biotechnology,* 13 : 686 - 691.

Neale, A.D., Wahleithner, J.A., Lund, M. Bonnett, H.T., Kelly, A., Meeks-Wagner, D.R., Peacock, W.J., and Dennis, E.S. 1990. Chitinase, β-1, 3-glucanase, osmotin and extensin are expressed in tobacco explants during flower formation. *The Plant Cell,* 2 : 673 - 684.

Sasaki, T. 1994. An overview of rice genome research program: three year progress report. Proceedings 7th Meeting of the International Program on Rice Biotechnology, Bali, Indonesia.

Tada, Y., Sakamoto, M. and Fujimer, T. 1990. Efficient gene introduction into rice by electroporation and analysis of transgenic plants : use of electroporation buffer lacking chloride ions. *Theor. Appl. Genet.*, 80 : 475-480.

Tansksley, S.D., Causse, M., Futton, T., Ahn, N. , Wang , Z. , Wu , K. , Xiao, J., Yu, Z., Second, G. and McCouch, S. 1992. A high density molecular map of the rice genome. *Rice Genet. Newsl.*, 9 : 111-115.

Uchimiya, H. and Toki, S. 1994. *Genetic engineering in rice plants. Improvements of cereal quality by genetic engineering,* ed. by R.J. Henry and J.A. Ronalds, Plenum Press, New York, 31-36.

Zhang, H.M., Yang, H., Rech, E.L., Golds, T.J., Davis, A.S., Mulligan, B.J., Cocking, E.C. and Davey, M.R. 1988. Transgenic rice plants produced by electroporation mediated plasmid uptake into protoplasts. *Plant Cell Rep.* 7: 379-384.

INDEX